ROUTLEDGE LIBRARY EDITIONS:
AGRICULTURE

Volume 15

PRIVATE AGRICULTURE IN THE SOVIET UNION

PRIVATE AGRICULTURE IN THE SOVIET UNION

STEFAN HEDLUND

WITH AN INTRODUCTION BY
ROY D. LAIRD

LONDON AND NEW YORK

First published in 1989 by Routledge

This edition first published in 2020
by Routledge
2 Park Square, Milton Park, Abingdon, Oxon OX14 4RN

and by Routledge
52 Vanderbilt Avenue, New York, NY 10017

Routledge is an imprint of the Taylor & Francis Group, an informa business

British Library Cataloguing in Publication Data
A catalogue record for this book is available from the British Library

ISBN: 978-0-367-24917-5 (Set)
ISBN: 978-0-429-32954-8 (Set) (ebk)
ISBN: 978-0-367-25132-1 (Volume 15) (hbk)
ISBN: 978-0-429-28642-1 (Volume 15) (ebk)

Publisher's Note
The publisher has gone to great lengths to ensure the quality of this reprint but points out that some imperfections in the original copies may be apparent.

Disclaimer
The publisher has made every effort to trace copyright holders and would welcome correspondence from those they have been unable to trace.

PRIVATE AGRICULTURE IN THE SOVIET UNION

Stefan Hedlund

With an introduction by Roy D. Laird

Routledge
London and New York

First published 1989 by Routledge
11 New Fetter Lane, London EC4P 4EE
29 West 35th Street, New York, NY 10001

Typeset by LaserScript, Mitcham, Surrey.

Printed and bound in Great Britain by
Mackays of Chatham PLC, Chatham, Kent

British Library Cataloguing in Publication Data

Hedlund, Stefan
Private agriculture in the Soviet Union.
1. Soviet Union. Agricultural industries
I. Title
338.1'0947

ISBN 0-415-03126-5

Library of Congress Cataloging in Publication Data is available.

Contents

Tables

Introduction

Stefan Hedlund rightly credits Karl-Eugen Wädekin's 1973 'magnum opus' *The Private Sector in Soviet Agriculture* as being a must in any attempt to understand the place of the private sector in Soviet agriculture. Although Hedlund's approach is quite different from Wädekin's, it is an important sequel to Wädekin's book. This work makes two highly significant contributions to furthering the understanding of the 'permanent crisis' that plagues Soviet agriculture. (1)Hedlund has updated our knowledge of the vital role of the private sector in Soviet agriculture. (2)While not neglecting the purely economic side of the problem, he is the first Western scholar to explore extensively the human causes behind the unsatisfactory performance of the socialist sector in the *kolkhozy* and *sovkhozy*. Naum Jasny, the great pioneer of Soviet agricultural studies entitled one of his seminal articles '*Kolkhozy*, the Achilles' Heel of the Soviet Regime'.[1] Hedlund could well have borrowed from Jasny and titled this work 'Private Plots, the Achilles' Heel of Perestroika in Soviet Agriculture'.

Significantly, Gorbachev and Hedlund both stress that any ultimate solution to the chronic Soviet agricultural problem must come primarily from a fundamental change in the mind-set of both the peasants and their rural managers, including the party leaders at the *raion* and *oblast* level. As the Brezhnev era of stagnation proved, merely pouring more rubles into the countryside while kowtowing to the local party prefects is no solution.

The Soviet agricultural problem is complex. Part of the problem is a peasantry that has been taught that 'sloppy work [in the socialist sector] is permissible'. Most ironically, as Hedlund documents, while the private plots remain an essential contributor to the food supply (and Gorbachev is encouraging greater private activity), much of the mentality that is associated with poor worker performance in the socialist sector is rooted in attitudes stemming from the persistence of the private sector. The Marxist–Leninists repeatedly proclaim that inevitable contradictions pose fatal flaws in 'capitalist' systems. Hedlund shows that major contradictions exist, and wreak their havoc, in the Soviet command system as well.

Again, the most important contribution of this work is that it goes beyond mere economic explanations. Hedlund documents that the Soviet agricultural system is a psychological 'black hole' that swallows and destroys all effort to achieve improvement, primarily because the system is designed for political control and not production – another source of major contradiction.

When the human side of the equation is accounted for, at best, Gorbachev's restructuring of the countryside seems destined to lead to a 'Pyrrhic victory'. Unless Gorbachev and his colleagues are prepared to make fundamental changes in the system itself, which they have emphatically rejected, achieving the imperative changes in attitude of both the workers and the managers seems impossible.

Roy D. Laird
Professor of Political Science
and
Soviet and East European Studies
The University of Kansas

Preface

In a certain sense, this book forms a sequel to a previous book of mine, entitled *Crisis in Soviet Agriculture*, which was published in 1984.[1] At the time, I received considerable criticism for my use of the word 'crisis', and in the sense that neither collapse nor noticeable recovery has occurred since then, my critics were of course right. The reason for returning now to that same subject, however, is not simply to seek justification for definitions used. Rather, it is due to a more personally felt need to try once more to understand how it is possible fort this colossus to continue its grossly inefficient existence, or perhaps even how it is possible for the Soviet leadership to *allow* it to do so.

Consequently, rather than just documenting the continued poor performance of Soviet agriculture in overall terms, I have tried to present a picture of the impact on the Soviet system as a whole of the stubborn refusal to implement any form of structurally important changes in agriculture. This statement of intent may perhaps seem a bit pretentious, but there is certainly no shortage of indications from the Soviet side that the problems at hand are of paramount importance. In his speech to the 27th Party Congress, in February 1986, Gorbachev stated that 'now, as never before, agriculture needs people who are prepared to work actively, with professionalism and an innovative bent'.[2] In a September 1987 decree from the Central Committee, it was stated quite unequivocally that agriculture is currently *the* pivotal task of Party policy.[3] In the April 1988 issue of *Novyi Mir*, the well-known reform economist Nikolai Shmelev writes that 'the crisis in our agriculture is obvious to all',[4] and at his May 1988 meeting with leading representatives of the Soviet mass media Gorbachev confesses that 'the food situation worries us. It worries us a lot.'[5]

Against this background, it is hardly surprising that shortly after his appointment as General Secretary, in March 1985, Gorbachev found that investment in agriculture had reached its 'optimal level'.[6] It is also fairly logical that the entire planned increase in agricultural output during the current (1986–90) five-year plan is to be the result of improved labour productivity.7 These latter statements in particular indicate quite clearly that after decades of priority for mechanization and large scale, it is now finally the human element, the *chelovecheskii faktor*, that is being placed in focus. As Gorbachev put it in his congress speech, 'the main driving force in the advancement of the agro-industrial complex, indeed its very soul, is, has been and will remain man.'

Success in this endeavour will obviously require not only specific training in new methods of work and management, but also a broad transformation of social and psychological attitudes and patterns of behaviour. It is symptomatic, for example, that the above-mentioned September decree concluded by calling upon Soviet mass media actively to propagate the experiences of *perestroika* in the agro-industrial complex.[8]

Given this conception of the problems at hand, the methodology to be used below is somewhat different from that of my previous book. There a chiefly institutional approach was used in order to disentangle some of the seeming absurdities and irrationalities that mark the behaviour of both the peasants and their 'prefects'. Here the main ambition is to investigate the moral and psychological impact on those people *inside* agriculture – peasants as well as managers – who for decades now have been taught that grossly inefficient management and sloppy work are quite permissible, and perhaps also on those *outside* agriculture who are increasingly being brought in to 'help out', and who are consequently able to develop private activities for private gain, while in the process of building socialism. Some parts of the argument pursued below will consequently be of an interpretative and at times perhaps even speculative nature. After all, when it comes to explaining human behaviour there is no solid truth, and thus also a limited amount of hard provable evidence.

My own impression – gained during work on this book – is that of a rather peculiar, and potentially dangerous, contradiction between the official talk about reform, restructuring and the speeding up of the economy, on the one hand, and the actual policy measures that have been implemented in the field of agriculture on the other. To my mind, the pronounced wager that is presently being placed on various forms of private initiative may serve not only to produce a further spread and strengthening of a form of *muzhik* ideology, which is the very antithesis to what Gorbachev needs for success with his programme of *perestroika*, but also to defuse somewhat the pressure for change in the socialized sector, as its private counterpart is urged into a continued covering up of some of its worst malfunctions. In this sense, the wager on the plots, and on various forms of contractual arrangements between the two sectors, may actually turn out to be doubly harmful.

The implications for the future of not only Soviet agriculture but of the Soviet economy at large are certainly rather gloomy. To put it quite simply, unless the behaviour of the Soviet peasants and their 'prefects' can be altered, there will be no *perestroika* in agriculture, and if there is no *perestroika* in agriculture then there will be no *perestroika* in the rest of the economy. The time that would realistically be required to complete this process has been set by the noted Soviet agricultural economist and academician Vladimir Tikhonov at about 25 years, or at a period corresponding to a change of at least two generations of Soviet leaders.[9] What the result might be if the reform attempt should fail has been suggested quite bluntly by Shmelev: 'For our agriculture, the current period is certainly a decisive one. If people's expectations once again (for the umpteenth time) should be put to shame, apathy may well become irreversible.'[10]

Shmelev, however, is far from alone in having an almost apocalyptic view on the current situation. In January 1987, Sergei Zalygin, prestigious editor of the prestigious reform journal *Novyi Mir*, wrote the following:

> And it is now quite clear, that if we do not manage to bring about order and democratization, within fifteen years we shall have been transformed into one of the ... poorest countries in the world. Thus we shall have squandered once and for all our truly amazing natural resources – nature itself – and in the process we shall ourselves perish.[11]

Perhaps the most worrying aspect of this apocalyptic perspective is that time appears to be running out fast. By May 1988 Zalygin had shortened his time horizon to less than half: 'It is now obvious to all that we have maybe another five-six years to go, then it will all be finished. Then we may speak of *perestroika*, or of anything we may please, but our resources will have been irrevocably destroyed.'[12] The topic at hand would thus seem to be of some considerable importance.

Work on this book has been both facilitated and frustrated by the existence of Karl-Eugen Wädekin's conclusive work *Privatproduzenten in der sowjetischen Landwirtschaft*.[13] It has been facilitated in the sense that there has always been an amply documented source to turn to for information. On the other hand, it has also been frustrating to note that at times little new could be added to the already existing volume. Needless to say, my debt to Wädekin's work is greater than is apparent from the footnotes. As it is, however, I have tried to focus on events during the past decade, which were not covered by Wädekin, and perhaps I have also raised somewhat different questions. Whether this is sufficient to warrant a new volume is of course up to the reader to determine.

Another factor which has provided a mixed blessing is the sudden onset of *glasnost*, which more or less overnight, has changed the previous problem of obtaining information into one of a persistent feeling of being unable to keep up with all the new material, both analytical and descriptive, that is now regularly appearing in Soviet papers and journals. On balance, the benefits are undoubtedly greater, but one should not underrate the trauma that is involved in deciding when to actually lay down the pen.

As a rule, it is customary at this point in a preface to express gratitude to friends and colleagues for helpful comments and criticisms, put forward in conversation or in relation to previous drafts. My case is certainly no exception to that rule. Roy Laird, Alec Nove and Karl-Eugen Wädekin have all read in full a previous version of the manuscript. The accumulated weight of their comments and criticisms has certainly contributed greatly to making the final version of this book a very different product from that which was originally circulated. My gratitude on this count is also extended to Harald Ståhlberg, an old friend and agronomist who has patiently sought to enlighten me on the intricacies of what makes crops grow and animals bear offspring. Any remaining errors, omissions or misunderstandings are of course the sole responsibility of the author.

Research for this book has also benefited greatly from the extensive Slavic collections in the Library of the University of Illinois at Champaign-Urbana, and from the highly professional assistance of its library staff. Their help is hereby gratefully acknowledged, as is the generosity of Göte Hansson, the Head of the Department of Economics at the University of Lund, for granting me the right to lengthy absences. Financial support making the latter possible has been provided by *Stiftelsen Lars Hiertas Minne* and *Stiftelsen för främjande av ekonomisk forskning vid Lunds Universitet*. Not being a native English speaker, I would like to offer a concluding word of thanks to Alan Harkess, for correcting my own at times very personal opinion on how this language should be written.

Lund, May 1988
Stefan Hedlund

Chapter one

A black hole in the Soviet economy

In Soviet agriculture, small has traditionally been very far from beautiful. The tens of millions of those tiny household plots that have received such fame in the West may have been sturdy, and in a certain limited sense even highly productive, but in the eyes of the Soviet authorities they have definitely not been beautiful. During the past decade, this situation has undergone a considerable change, in the direction of a more open recognition of the important contributions that are being made by the plots towards national food supply. Officially, this change in attitude has been demonstrated by a number of decrees and pronouncements calling for increased support to be given to the private sector. During the last couple of years in particular, there has also been a quite pronounced ideological reorientation in matters concerning private agricultural activities. The contents and possible implications of this change in the course of Soviet agricultural policy will be an important topic for investigation in this book. Before proceeding to that task, however, we should perhaps underline that there is an obvious difference between a change in official policy and a change in the mentality and attitudes of those who are charged with implementing that new policy.

A central task to be undertaken below will thus be to penetrate beyond the official façade, and in so doing we shall have to delve into a partly unknown realm of Soviet agriculture. All those various reasons that – in the past at least – have combined to form a rather hostile official attitude towards the private plots have of course also left a firm imprint on the availability of information – qualitative as well as quantitative. The scattered nature of the available information is clearly manifested in the Soviet writings on the subject, and this problem will unfortunately also accompany us on the following pages.

The reasons behind these difficulties are of several kinds, the most important perhaps being that some of the activities in the private sector take place on or beyond the limits of the law, and can thus not be reflected in official data. Other reasons stem from the fact that we are dealing with many millions of small-scale producers who are operating on a part-time household basis, with make-shift resource inputs, and are often located in remote rural areas. The very nature of the private plot as a producing entity thus creates obstacles to statistical description which would be of considerable magnitude even in a society not marked by the traditional Soviet (i.e. pre-*glasnost*) attitude towards the creation and dissemination of information.

There are of course many different aspects to the peculiar and rather uneasy coexistence of very small-scale private and very large-scale socialized agricultural production. In this study only certain of these aspects will be dealt with. In order to illustrate which areas will be in focus, the following excerpt from Gregory Grossman's preface to the English edition of Karl-Eugen Wädekin's *magnum opus* on the private sector in Soviet agriculture is quoted:

> The private sector is of course an anomaly in the socialized, centralized, planned economy of the USSR. Economically it is backward, ideologically it is alien, politically it is suspect, and morally it stands in the way of the creation of the new socialist and communist man.[1]

What is really going on here? Against the background of Grossman's characterization, it would seem hard indeed to explain the existence of the private plots in terms that are remotely rational. Yet, it will be a main theme of this study that, in addition to the obvious function of providing a substantial share of the country's food supply, the plots also fill an important function as a form of 'political stabilizer'. By taking some pressure off the poorly functioning official system of production and distribution of foodstuffs, they serve to dampen a potentially dangerous growth of popular discontent, and by providing a 'private' refuge of activities outside the state-controlled sphere, they serve to defuse some latent discontent.

Although perhaps not consciously implemented, in a certain limited sense the policy of allowing such private agricultural activities can thus be considered a success, economically as well as politically. In a broader sense, however, it will be argued that this success has important elements of a 'Pyrrhic victory', where the moral and psychological impact on the individuals concerned may be of such a nature that it actually blocks the current attempts at 'restructuring' agriculture. In this sense, the 'support' that is rendered by the private plots may paradoxically serve to condemn the official, socialized sector of Soviet agriculture to remain in that state of 'permanent crisis', which was diagnosed by Roy Laird and Edward Crowley as early as 1965.[2] The discussion starts by examining the performance of this sector, in order to understand better how the need for private support has emerged.

A Permanent Crisis[3]

Ever since Marx and Engels made their celebrated statement – in the *Communist Manifesto* – on how capitalist development had freed a large section of humanity from the 'idiocy of rural life',[4] agriculture and what the Soviets refer to as 'real socialism' have given a distinct impression of being uneasy bedfellows. Several areas in Eastern Europe that were once surplus producers and large grain exporters – notably Poland and Prussia – have been transformed into net importers. The Soviet Union itself is certainly not an exception in this respect.

Seven decades after the Bolsheviks' October *coup d'état*, and more than half a century after Stalin's mass collectivization of the Soviet peasantry, agriculture remains a major headache to the Soviet economy. It is symptomatic, for example, that one of the few things on which there seems to be a broad consensus among the proponents of Mikhail Gorbachev's current policy of *perestroika* is that any attempt at change, if it is to be successful, must start by approaching the problems in agriculture. The process that has placed the Soviet superpower in this rather embarrassing position bears a heavy imprint of those largely *ad hoc* policy measures that were once deployed by Stalin, in his private war against the peasantry. Political ambitions for power and security produced an agricultural policy that placed *extraction* before *production*,[5] and the consequences are still there to be observed. In 1974, for example, Moshe Lewin pronounced the following verdict:

> Soviet agriculture has not yet managed to effect a real technological revolution similar to the one which took place some time ago in other developed countries. Agriculture is still rather primitive and a great problem and there is no doubt that the consequences of the first quarter of a century of *kolkhoz* history still weigh heavily and are far from having been definitely overcome.[6]

It is perhaps not so hard to understand that such a policy was pursued during the first troublesome decades of Soviet power. The fact, however, that the post-Stalin period has failed to produce any significant structural alterations of the *kolkhoz* system is perhaps in somewhat greater need of elucidation. It is rather tempting, for example, to wonder whether perhaps there are some important structural features of the overall political and economic system that conflict with the highly specific demands of agriculture's largely biological mode of production.

To answer that question, we shall invoke the voice of Alexander Yanov, a former Soviet journalist and a prolific writer on the attempts at agricultural reform during the Khrushchev era, who is presently living and working in the United States. Presumably, he has been asked the very same question on numerous occasions, and the following may well have been his answer:

> Inevitably the answer to the question 'Does the system work?' depends on what one means by 'work'. If it refers to political control, then the *kolkhoz* system works very well; if it refers to food production, then the system does not work, for it was not designed to.[7]

Their stubborn refusal to implement any form of important changes in the structure and operation of agriculture has forced Soviet policymakers instead to commit an ever-increasing amount of resources to that sector, simply in order to prevent, or at least postpone, the seemingly inevitable need to decide eventually on a radical change. In this sense, Soviet agriculture has come to assume the image of one of astronomy's Black Holes, capable of absorbing whatever resources come near whilst allowing very little to trickle out at the other end. It will be the purpose of the following section of this chapter to take a somewhat closer look at the volume of resources that has been thus absorbed.

A Black Hole

In all fairness, we should perhaps start by noting that if no considerations are made of the costs involved, the performance of Soviet agriculture in the post-Stalin era does present a rather impressive picture. Table 1.1 shows that the all-important output of grain, for example, increased by 48.2 per cent over the period as a whole, or by 68.7 per cent if the highly unfavourable years 1981–85 are excluded. With the exception of potatoes, the other products in the table show an even more impressive pattern of development, with the nutritionally vital production of vegetables almost doubling (over the whole period), while output of meat more than doubled, and eggs more than trebled.

It may be added here that the relatively less successful performance of grain production is of particular importance in the light of the rapid expansion of livestock production. It is a rather well-known fact that Soviet feed balances have long included excessive amounts of

Table 1.1 Soviet agricultural output, 1956–85
(million tons; eggs: million pieces; annual averages)

	1956–60	*1961–65*	*1966–70*	*1971–75*	*1976–80*	*1981–85*
Grain	121.5	130.3	167.6	181.6	205.0	180.3
Vegetables	15.1	16.9	19.5	23.0	26.3	29.2
Potatoes	88.3	81.6	94.8	89.8	82.6	78.4
Meat	7.9	9.3	11.6	14.0	14.8	16.2
Milk	57.2	64.7	80.6	87.4	92.7	94.6
Eggs	23.6	28.7	35.8	51.4	63.1	74.4

Source: Narkhoz (1986), pp. 180–1.

concentrates, in relation to roughages, and that this lack of balance has resulted in very poor results in terms of both milk yields and animal weight gain. Since, moreover, the bulk of concentrate feed has tended to be grain, an expansion of the livestock sector has automatically added extra pressure on grain production.[8] It is consequently not surprising that much of Soviet grain imports during recent years has been intended for feed purposes.

This rather favourable overall picture needs, however, to be qualified in a number of different respects. First of all, if the all-important grain is used as an example a rapidly growing gap is found between planned and actual output. In 1966–70, the plan was only just fulfilled, but then the troubles started. In 1971–75, the total deficit between planned and actual grain production amounted to 67 million tons. In 1976–80, 75 million tons was recorded, and in 1981–85 it came to no less than 300 million tons. The accumulated deficit for the period 1970–85 thus amounts to a staggering 442 million tons, or roughly equal to two and a half times the average annual harvest in 1981–85. With harvests of 210 and 212 million tons respectively for 1986 and 1987, a further 84 million tons were added to that deficit.[9]

On the seemingly reasonable assumption that official plan targets reflect what is considered a reasonable return to allocated resources, these figures present a gloomy picture indeed. Moreover, when the fact that the Soviet population grew from 197.9 million in 1956, to 276.3 million in 1985, or by 39.6 per cent (by 33.6 per cent over the period 1956–80), is taken into account, the picture is hardly made any brighter.[10] Grain production measured *per capita* did increase from 0.58 tons in 1956–60, to 0.72 tons in 1971–75, but for 1976–80 the attained level was no more than 0.78 tons, and for 1981–85 an absolute decline was registered, to 0.66 tons.[11] Khrushchev's old promise of reaching an output of one ton of grain *per capita* of the Soviet population is thus being reduced to a rather hollow utopia.

Overall performance in aggregate output terms, however, describes only one aspect of Soviet agriculture. The real seriousness of the situation is brought out only when we consider the volume of resources that has actually been consumed in the process. Let us therefore look in somewhat greater detail at the allocation to agriculture of land, labour and capital, the three 'textbook' factors of production, starting with land.

Land Use

One of the more spectacular achievements of the 'Soviet Model' that has been recorded during the seven decades of Soviet development, is undoubtedly that of the Virgin Lands campaign of the 1950s. At the time of Stalin's death, Soviet agriculture was marked by serious stagnation. The solution chosen by Khrushchev was to add new land in order to achieve a rapid increase in output. A campaign was launched, in traditional Soviet fashion, with considerable pressure from the centre and with substantial political prestige involved for local officials. Viewed from the input side, the results were nothing short of stunning.[12]

The campaign focused mainly on the idle steppelands of northern Kazakhstan, and the pictorial evidence alone, of rows of combine harvesters driving abreast from horizon to horizon, is indeed rather impressive. During 1954, a total of more than 17 million hectares was ploughed up, and in the following year a figure of close to 30 million hectares was reached. Subsequently, progress was somewhat slower. In 1960, however, a record total of some 41 million hectares was achieved.[13] For the sake of comparison, we may note that this area corresponds in size to the entire cultivated area of Canada, or to that of France, the UK, West Germany, Belgium and Scandinavia taken together.

By 1960, however, the outer limit had been reached for what could be brought under cultivation without major investment in land improvement. Further attempts at expansion were frustrated by the dropping out of marginal lands. In the period 1950–60, the total sown area in the Soviet Union increased by almost 40 per cent, from 146 to 203 million hectares, and hence the option of increasing agricultural production by means of adding new land had been largely eliminated from the agenda. In 1986, following massive investments in land improvement during the Brezhnev era, the total area stood at merely 228 million hectares.[14]

Capital Investment

With the reserves of easily cultivable land thus being largely exhausted, it is hardly surprising that a massive programme of capital

Table 1.2 Capital-output relations in *kolkhozy*, 1965–80 (billion 1973 rubles)

	1965	*1970*	*1975*	*1980*
Capital stock	42.3	60.0	91.7	109.8
Output	35.4	42.3	42.0	41.5
Capital/output	1.19	1.43	2.17	2.63

Source: Narkhoz (1981), p. 254.

investment, in mechanization as well as in land improvement, was to be Brezhnev's policy. In the period 1965–80, investment in Soviet agriculture increased from 14.4 to 41.6 billion rubles – or by almost 200 per cent – bringing agriculture's share of total investment in the Soviet economy from 22 to 28 percent.[15] In comparison with other industrialized nations, both the *level* and the *direction* of Soviet agricultural investment can thus be seen to show unfavourable patterns.

Economic development is normally interpreted as increasing productivity in the primary sector (agriculture and extraction), which allows a transfer of resources (initially labour) into the *secondary* (manufacturing) and eventually the *tertiary* (services) sectors. In this perspective, the Soviet case currently stands out quite distinctly as one of a *de facto* reversal of industrialization. Instead of providing fuel for expansion of the secondary and tertiary sectors, the primary sector is acting as a drag on development, by actually *increasing* its claims on available resources, in relative as well as in absolute terms. It would require considerably more space than is available here to provide an explanation of how this pattern of development can possibly be combined with the persistence of large grain imports.[16] A brief look at Table 1.2, however, may provide a general impression of the forces that are at play.

In spite of an increase in the capital stock by 160 per cent, over the period 1965–80, output in the *kolkhoz* sector grew by merely 17 per cent.[17] For the farm sector as a whole, the marginal investment needed to produce an extra ruble's worth of output has quadrupled since the early 1970s, and is now three times higher than in industry.[18] Sombre as this may seem, however, the problem is not only one of a declining efficiency of investment. Even the achieved level ought to give cause for alarm. As late as 1985, official Soviet sources characterized less than a quarter of all agricultural work as 'mechanized',[19] which means that many millions of Soviet peasants are still engaged in heavy manual labour, aided by simple and often make-shift tools and implements.

Table 1.3 Share of the rural labour force in total, 1950–85 (per cent)

	1950	*1970*	*1975*	*1980*	*1985*
United States	12.1	4.5	4.1	3.6	3.1
Soviet Union	53.9	32.2	28.8	26.4	25.4
Japan	51.6	17.4	12.7	10.4	8.7
Italy	43.9	18.2	15.2	13.2	11.2

Source: Goodman, Hughes and Schroeder (1987), p. 102.

One possible explanation might of course be that the use of other factors had fallen, but this has hardly been the case. Both land and labour resources are still being pushed to their respective limits, and we are thus left with the impression of the Black Hole that was indicated above. Yanov certainly does not mince his words in concluding that the *kolkhoz* is going bankrupt: 'It turns out that money is being spent without results. Increasing the complexity of the operation leads to its decline. The *kolkhoz* seems to be a bottomless pit.'[20]

Labour Force

Labour use is perhaps the most baffling of all the indicators of poor productivity in Soviet agriculture. Soviet sources normally estimate the productivity of the farm labour force at 20–25 per cent of the US level (a favourite Soviet yardstick), while Western sources put it at no more than 10 percent.[21] Estimates based on the Soviet 1970 census, which have been presented by D. Gale Johnson and Karen Brooks, also show that the number of able-bodied workers per hectare is far greater in the Soviet than in the US case.[22] Consequently, it is hardly surprising to find that compared with other industrialized nations, the Soviet agricultural labour force has not only been declining slowly but also remains at a very high level (Table 1.3).

Substantial as these figures may seem, however, they still fall short of reflecting the true magnitude of Soviet agriculture's claim on human resources. It is a rather well-known fact, for example, that every year at harvest time the Soviet countryside is invaded by millions of urban residents who are drafted into such work. Official statistics do report the number of people who are thus 'attached' to agriculture, but only in terms of aggregated man-years. This figure has risen steadily, from 0.5 million in 1960, to 1.0 million in 1975 and 1.4 million in 1985.[23] Thus presented, its importance appears to be fairly small, corresponding in the latter year to an increase in the annual *kolkhoz* and *sovkhoz* labour force by no more than 5–6 per cent.

In order to find out how many individuals are actually involved, and what the actual increase is at harvest time, we need to know what conversion factor is used in order to produce man-years. Here we are helped by a Soviet source from 1981, which claims that in 1978 a total of 15.6 million urban residents, around half of whom were industrial workers, spent about a month each on the farms, 'helping out' with the harvest. This figure was reportedly 2.4 times greater than in 1970.[24] If we apply this factor 12 to our official annual figures, we find that the number of people involved increased from 7.2 million in 1970, to 15.6 million in 1980. The difference between the two sources may result from the fact that the latter only includes 'otherwise employed' persons, thus excluding for example large numbers of students. Since it is not known how many 'real' peasants are working at harvest time it is not possible to calculate what increase is actually provided, but it might conceivably be well over 50 per cent. Chapter 6 discusses at greater length the type of 'help' that is thus provided, and the consequences for industry of being subjected to this form of resource 'taxation'.

These developments suggest that poor labour productivity is the real stumbling block to Soviet agricultural development. This impression is consistent with evidence presented by Yanov on the attempts made during the 1960s to introduce a system of self-management and payment by results which, if properly implemented, would allegedly have made millions of unskilled peasants redundant.[25] It is also consistent with Gorbachev's current claim, that a successful *perestroika* would free 10 million people from agriculture.[26] All of this indicates rather clearly the importance of the human element, an issue which certainly lies at the very heart of the presentation in this book. With this, let us proceed to see what the Soviets have been prepared to do.

Instead of a Cure

As the 1970s wore on, it must have become gradually and painfully obvious to increasing numbers of the Soviet leadership that something would simply have to be done about the problems in agriculture. The outcome was the adoption of a 'Food Programme' – the *Prodovolst vennaya programma* – to cover the period 1982–90. Originally commissioned in 1978, this programme was adopted at the May 1982 Central Committee plenum.[27] Whatever doubts and/or opposition may have existed against its launching must certainly have been swept away by the disastrous 1981 harvest.[28] With the benefit of hindsight, we can say that this programme recalls the celebrated quotation from Horace's *De arte poetica*, about the mountains being pregnant but succeeding only in giving birth to a ridiculous little mouse: '*Parturiunt montes, et nascetur ridiculus mus.*' To arrive at such a conclusion, however, would have required neither hindsight nor a knowledge of the classics. Already at the time of its inception, critical voices were heard – in the West at least – questioning whether the new programme would represent a fundamentally new policy or simply 'more rhetoric'.[29] As we now know, reality would bear these voices out.

Although much ado has been made of the Soviet claims that the 1986 and 1987 grain harvests reached the surprisingly high levels of 210 and 212 million tons respectively, it remains a fact that the average for 1981–85 was no higher than 180.3 million, whereas the target for that period was 238–243 million tons. The original target level of 250–255 million tons for 1986–90 is probably nothing but pure utopia.[30] Indeed, as Wädekin has noted, the guidelines for the 1986–90 plan sets a target of 250–255 million tons for the year 1990, rather than for the annual average. As the latter would have implied 280 million tons for 1990, ambitions have been reduced quite considerably from the original euphoria of the Food Programme.[31] Even this scaled-down target, however, may well prove to be overambitious. It should also be noted that several Western observers have questioned – with varying degrees of caution – the reliability of the 1986 and 1987 figures.[32]

Against this background, the Food Programme can be seen as a splendid manifestation of the unwillingness – or inability – of the decaying Brezhnev leadership to face up to the real needs of the situation. It has even been argued, by Stephen White, that the programme may actually have been a response to an increasing flood of letters, highly critical of the food situation, that had been received by Soviet media and by official Party and state organs.[33] The vigorous launching of a specialized new programme might well have been intended to persuade the population that its leaders had at least a *determination* to do something, and thus perhaps help ease some of the growing pressure of popular discontent with the food situation.

The actual contents of the programme clearly illustrate the intentions of 'purchasing' yet another postponement. Any discussion of real change is conspicuous by its absence, while the readiness to sustain the already substantial volume of resources poured into agriculture is conspicuous by its sheer magnitude. The 'agro-industrial complex' as a whole, which, as pointed out above, is a broader definition than *po vsemu kompleksu rabot*, would continue to receive around a third of the total investment in the Soviet economy. Furthermore, state procurement prices were to increase at an estimated annual cost of 16 billion rubles, while investment in the 'socio-cultural' sphere (housing, schools, kindergartens, etc.) was to

Table 1.4 *Kolkhoz* investment and debt, 1965–80 (billion current rubles)

	1965	*1970*	*1975*	*1980*
Investment	4.4	6.6	9.2	10.3
New loans	1.4	2.2	3.3	4.7
Short term debt	0.3	2.4	10.1	25.7
Long term debt	3.9	10.3	17.8	34.0

Source: Narkhoz (1981), pp. 341, 528, 531.

receive another 3.3 billion rubles annually. Finally, it was decided to write off a total of 9.7 billion rubles in outstanding debt, and to reschedule a further 11 billion.[34]

The latter measures are of particular importance, in that they illustrate the almost resigned attitude that was taken to matters of financial discipline. The relation between investment and output has been described above, and here we may add the consequences of a decision taken in 1966, to make wages a priority expenditure. While *kolkhoz* income (net of purchases of current material inputs) in the period 1970–80 actually *decreased*, from 22.8 to 19.6 billion rubles, total wage payments *increased* from 15.0 to 18.6 billion, thus increasing the share of wages from 66 to 96 per cent of total income.[35] Table 1.4 illustrates the role of credits in *kolkhozy* during the period leading up to the Food Program.

This shows a text-book example of what János Kornai has referred to as a 'softening' of the budget constraint.[36] For decades, *kolkhoz* chairmen and *sovkhoz* directors have been taught by everyday experience that poor financial discipline is permissible, that losses will be covered via subsidies and that failure to repay credits will be rewarded by the issuing of more credits. It is against this background that the current Soviet statements about shifting from 'administrative' to 'economic' methods of management must be viewed. We shall have call in the concluding chapter of the study to return at greater length to this important issue.

Again, we are thus faced with the impression of a Black Hole, of a seemingly unlimited capacity to absorb new resources without yielding much improvement in output. It will be argued in the remainder of this study that the last resource to be sucked into this Hole is that of private agricultural activities. While it is doubtful whether this mobilization of private initiative will contribute much more than merely to postpone the inevitable need finally to undertake serious reform, the consequences of this last rescue operation may prove to be considerably more costly than those of any of its predecessors. To substantiate this claim shall be the main ambition of the study. Let us start with a brief look at how the presentation will be organized.

Plan of the Study

We shall begin our discussion in Chapter 2, by presenting a picture of the emergence of a private sector that is made up of various forms of auxiliary agricultural activities, all of which share one common trait in serving as a vital support for the official, socialized large-scale sector. This presentation of a 'support' sector continues in Chapter 3, by looking more closely at two 'fringes' of auxiliary agricultural activities – one official and one private – that have recently come to assume a rapidly growing importance. The combined impression of these two chapters is that private agriculture is tolerated simply because it is economically necessary.

The attitudes of Soviet officialdom to such activities are the topic of Chapter 4, which starts by looking at a notorious bone of contention, in the form of the long ranging debate over whether we are dealing here with 'private' or 'personal' activities. Out of this debate two different strands of attitudes are extracted. On the one hand is the ideological aspect of allowing activities for private gain, while in the midst of building socialism and creating 'Soviet Man'. The other strand is represented by popular discontent with alleged 'speculation', on the plots and above all on the *kolkhoz* markets. These two dimensions will be dealt with in turn. As a result of the discussion will follow a more complicated picture, comprising the moral and political harm caused by the private sector as well as its perceived economic necessity.

This picture of at times openly hostile official attitudes towards the economic necessity of allowing such politically and morally suspect activities, will in turn serve as background to the presentation, in Chapter 5, of the practical manifestations of policy measures which will be seen to range from reluctant support to active harassment. This chapter will serve as a bridge between the attitudes that are held by officials and the reality that confronts the peasants.

Chapter 6 presents the other end of the spectrum, by looking at the role of the private sector from the point of view of the individuals directly concerned. Here attention is focused on the situation of the individual in a generally repressive setting, in order to illustrate the consequences for the system as a whole that result from the political and moral aspects of the function of plot activities as a political stabilizer.

In the concluding chapter, Chapter 7, it is argued that these stabilizing mechanisms are composed of elements that may present serious obstacles in the way of current attempts at economic reform, or 'restructuring'. Success along the path of *perestroika* will in this perspective be seen to require suppressing some of those mechanisms which for decades have provided the basic stability of the Soviet economic and political system. This may well prove to be a dangerous balancing act.

Chapter two

The emergence of a 'support' agriculture

A discussion of the private sector in Soviet agriculture must begin with the realization that the picture is rather kaleidoscopic. First of all, not all of those activities that take place outside the official sector proper, i.e. outside the state (*sovkhoz*) and collective (*kolkhoz*) farms, are 'private', even in a broad sense of the word. The notoriously poor performance of the 'real' farms, that was referred to above, and perhaps even more so of the official system of procurement and distribution of foodstuffs,[1] has led to the emergence of a food policy under which everyone must increasingly fend for himself. One manifestation of this policy is the development of a system of ancillary farms at state enterprises and institutions, factory farms that have come to form a separate category of their own. These have been included here because they form part of the auxiliary sector (*podsobnoe khozyaistvo*) of Soviet agriculture, and thus share some important features with the private plots.

Second, for the remainder of 'proper' plots, the Soviets are very careful in using the word 'personal' (*lichnoe*) instead of 'private' (*chastnoe*). Thus, the official term for the private sector is 'personal auxiliary agriculture' (*lichnoe podsobnoe khozyaistvo*), normally referred to under the abbreviation LPKh, and as shown below, the distinction is not entirely a matter of linguistic gymnastics. A number of good reasons can be (and are) advanced for distinguishing between Soviet 'personal' agriculture and 'private' agriculture in its Western sense. These will receive special treatment in Chapter 4. Otherwise, however, for reasons of common usage, the term 'private' will be employed (except in direct quotations from Soviet sources).

Third, it is a rather common misconception that plots should be available only to people who live and work on the farms. In practice, a contrary development is actually observed, in that rapidly growing numbers of the urban population are taking up plot activities, while the share of the farm population amongst those who have access to plots is shrinking (measured in household numbers, if not in area, since the average size of plots in the former category is considerably smaller than the latter). This rather peculiar process will receive attention throughout this study, but Chapter 3 will be specifically assigned to look more closely at the composition and the conditions of existence on this private 'fringe'.

Those plots that are placed at the disposal of members of *kolkhozy* and of workers and employees in *sovkhozy* and other state enterprises and institutions in rural areas are now examined. These form the real core of the private sector, and will be referred to below as the private sector 'narrowly' defined.

The Origin of the Plots

The origin of the private plots provides an important clue to the puzzlement expressed above, as to their present role, and may thus merit some special attention. From the impressive works of Robert Conquest and Moshe Lewin,[2] we are familiar with the horror story of Soviet mass collectivization, when the introduction of a 'second serfdom'[3] deprived the peasantry of the rights to their land, their animals and even their tools and implements. Against the background of these – mildly speaking – shocking accounts, it would seem fair to assume that rights to small household plots and minimal holdings of domestic animals were granted as a retreat from 'full collectivization', simply in order to provide the peasantry with a chance for survival. Such an interpretation is also consistent with the prehistory of private agriculture under Soviet power.

In a broad historical analysis of the role of the private sector, the Estonian economist Ivar Raig notes that 'the real history of regulation of personal auxiliary agriculture started in 1930, with the drafting of the new charter of agricultural *artely*.'[4] In a limited sense, this is of course true, as forced collectivization implied that the questions of private plots and private livestock holdings would have to be resolved. There is, however, also an important qualification to be added, in that there was still not any clear official stand on what the actual rules of the game should be.

While a model charter was indeed formally ratified by the First Congress of *Kolkhoz* Shock Workers, convened in March 1930, this was far from the final word on the issue. In the coming years a heated debate ensued, where Stalin presented himself in a 'liberal' position, in defence of the peasants against Party hard-liners calling for quite severe restrictions.[5] Given the practical circumstances of the time, the 1930 charter was clearly influenced by the traditional nature of the peasant household. It is thus natural that the issue of private livestock holdings in particular would prove to be controversial, causing much trouble and even riots. There were, however, other problems as well, all of which served to delay a final settlement. Lewin summarizes these ambivalences and their outcome in the following way:

> By 1935 the government seemed to have learned the lesson and have reached a decision to compromise with the peasants by legalizing the plot fully and guaranteeing in the new status of the *kolkhoz* the right to a certain number of cattle to be kept on the family farm.[6]

The outcome was a revised model charter of agricultural *artely*, which was ratified by the Second Congress of *Kolkhoz* Shock Workers, convened in February 1935. Once adopted, this new charter came to serve as the legal framework of the *kolkhoz*, which at the time was referred to as *artel*. It would remain in force unchanged up until 1969, when a second revision was undertaken, to be ratified by a Third Congress of *Kolkhoz* Shock Workers.

Although the first revision of the 1930 charter implied no major changes in principle, the very process of revision in itself illustrates the controversial nature of some of the basic issues involved. At the sixteenth Party congress, held in June 1930, the following interesting statement could be heard:

> To demand that peasants upon entering the *artel* should immediately give up all forms of individual habits and interests, refrain from conducting activities in support of the official sector (cows, sheep, fowl, household gardens), and give up the possibility of extra income, etc. – would mean forgetting the ABC of marxism–leninism.[7]

This stand should be seen against the background of the prevailing attitude during the very first years of Soviet power – the period of War Communism – which had held that private activities in agriculture were to 'die out'. In a 1919 decree on the organization of socialist agriculture, for example, it had been stated that 'Nobody shall have the right to introduce private animals, birds or household gardens', and in a 1920 model charter of agricultural *artely* (a predecessor to the 1930 one), a 'full liquidation' of all private activities by *artel* members had been called for.[8]

What happened in the immediate aftermath of the worst excesses of collectivization and dekulakization has been characterized by Gelii Shmelev, the leading Soviet authority on the private sector, as a 'liquidation of cowlessness' amongst the *kolkhoz* peasantry.[9] Behind the euphemism, of course, is hidden that massive slaughtering of livestock which was the peasant response to being forcibly collectivized, and the new policy is certainly best interpreted as payment on debts incurred during the previous years.

In March 1932, a decree was issued condemning forced socialization of cattle, and those guilty of violating peasant rights were severely censured. A number of additional decrees followed, calling for sales of cattle to the peasants and for the provision of credits to those who could not afford to pay, all in an attempt to repair some of the previous destruction of livestock herds. During 1934 alone, almost 1.5 million calves were sold to peasants who had no livestock, and a total of over 100 million rubles was paid out in credits for such purposes.[10] According to Raig, in the period 1932–38 private livestock holdings increased from 10.2 million head to 25.1 million, thus reducing the share of *kolkhoz* peasants without livestock to only 18.9 per cent.[11] Diplomatically, nothing is said about what happened in the years 1930–32, nor shall we have reason here to deal further with these tragic events.[12]

The policy of supporting the socialized sector via a wager on private initiative continued throughout the 1930s, including the extension – in 1933 – of plot rights to workers living in rural areas, for use in their free time.[13] This development illustrates clearly the core of a problem that still remains unresolved. On the one hand, the private sector made a valuable contribution towards food supply, but on the other it also provided rapidly rising incomes for the peasants. Logically, in 1939 a 'struggle against violations of the *kolkhoz* charter' followed, as a result of which about 40 per cent of the *kolkhoz* households were deprived of 'surplus' land holdings, scattered strips of land which once 'returned' were often of little (if any) use to the *kolkhoz*. In commenting on such practices, Shmelev makes the following important observation: 'In the majority of *kolkhozy* such normally high-yielding lands became idle and overgrown with weeds.'[14] Raig's comment is also typical of subsequent policy: 'However, implementation of the decree often took place dogmatically, without concern for local conditions.'[15]

This zig-zag course of policy towards the private sector has continued up to the present. During the war, restrictions were eased and production increased. After the war a new attack on 'breaches of *kolkhoz* discipline' was made, which reduced the sector's importance. In 1953, compulsory procurement quotas for the plots were reduced, and an upswing followed. In 1957, finally, it was decided to abolish such quotas altogether, effective as of January 1958.[16]

All these retreats and advances share one important trait. They illustrate rather forcefully on the one hand the seemingly instinctive distrust and dislike of private plot activities that existed (and continue to exist) amongst Soviet officialdom, and on the other a grudging and periodically suppressed realization of dependence on the private sector's contribution to total agricultural output. Given the importance that will be placed below on the attitudes of Soviet officials to the private sector, we shall digress briefly here to look somewhat more closely at the consequences of the attacks on the plots that were initiated by Khrushchev in the late 1950s. The turbulence that marked events during this period will serve to reveal some important aspects of the role and conditions of existence of the private sector. The lessons drawn from this period ought of course to be as important to Soviet policy makers of today, as they are for our understanding of subsequent policy measures.

Khrushchev's Attacks

The very mention of the words 'attack' or 'campaign', that are normally used with reference to Khrushchev's policy towards the private sector, almost instinctively makes one look for a major political occasion to be used as a starting point, and it is hardly surprising that some sources have associated the change in policy (for a change it was) with the December 1958 Central Committee plenum.[17] As Wädekin has shown, however, the starting point would be more correctly placed a couple of years before, amidst the general euphoria of the 1956 bumper harvest and the subsequent near-classic boasts about catching up and overtaking the United States in the *per capita* production of milk – by 1958 – and of meat – by 1960 or 1961. The distinction is important, in that the latter version presents the subsequent policy against the plots as a logical consequence of the belief that the socialized sector had reached a level of development where it was strong enough to shoulder the responsibilities of the private sector.[18] As we shall see, however, reality would show this belief to be quite unfounded.

A First Advance

The years immediatedly preceding Khrushchev's first attack had been marked by a rapid growth in the importance of the private sector, and in a sense the change in policy can be seen to have grown out of various reactions against this development that were voiced *inter alia* at the twentieth Party congress, held in February 1956. As is well known, however, this congress was almost totally dominated by the traumas of de-stalinization, and it would only be in the following months that a number of decrees were issued, calling for specific measures to be taken. Of particular importance in this context is a decree of 6 March, entitled 'On Monthly Advances to *kolkhozniki* and

Monthly Labour Payments in *Kolkhozy*'. As Wädekin has put it, 'this directive bore the seed of all subsequent measures against the private sector – measures which were to be applied with particular emphasis during periods of economic hardship and especially after poor harvests.'[19]

The basic thrust of the 6 March decree was aimed at strengthening the role of the socialized sector as a source of income for the *kolkhoz* households, and as a general policy goal this was surely laudable. If the vision of the future for the Soviet Union was that of a modern, industrialized superpower, on a par with the United States, there would then surely be no room for a medieval-type, small-scale agricultural sector, in the hands of peasant families who depend on the labour of pensioners, invalids and children. We shall have repeated reasons below to return to this important issue.

The problem, however, was that the decree could be implemented in two very different ways. Either the socialized sector could be strengthened, whilst allowing unchanged or even increasing incomes from the private sector, or pressure could be brought to bear on the private sector, so that the socialized sector could grow in *relative* importance, without any actual improvement taking place. As the former would be difficult indeed to achieve without a transfer of resources from the plots, it would in practice be the latter option that was favoured, and the consequences are there to be observed.

In the three years from November 1955 to November 1958, the total area of private plots in *kolkhozy* declined from 6.9 to 6.1 million hectares. Measured in terms of sown area, the reduction was smaller, but this seems to have been due to a more intensive cultivation.[20] Moreover, on looking at private livestock holdings, which is a better measure of the 'health' of the private sector, an even stronger impression of decline is obtained. In the twelve months from October 1955 to October 1956 alone, the number of cows dropped by more than ten per cent and pigs by close to a quarter.[21]

However important this decline may have been, the events in 1956–58 were only a prelude to what was to come. Even the seemingly beneficial act of removing compulsory delivery quotas from the plots would become part of the ambition to strengthen the socialized sector. As Wädekin points out, it was only by abstaining from these private deliveries that the state could afford 'morally and materially' to carry out the envisaged restrictions.[22]

A Full-Scale Attack

The signal for a full-scale attack came in December 1958, when Khrushchev declared that the *sovkhoz* sector, i.e. the state farms, had become strong enough to 'satisfy the requirements of their workers and their families for agricultural products', and furthermore that the 'existence of large plots and of livestock in personal ownership has become a serious impediment on the road to the further development of *sovkhoz* production.'[23] (As shown below, this focus on the *sovkhoz* would have a significance all of its own.) The necessary orders were promptly passed by the Central Committee.

By far the most important ingredient in Khrushchev's policy of cutting the private sector down to size was aimed at its very life nerve, the private holdings of livestock. In its official version, the new policy would rely on voluntariness, and on convincing the peasantry that by selling their cattle to the *kolkhoz* they would help increase overall efficiency, thus making everyone better off. Promises were also made that after having sold their cows, the peasants would be provided with milk from the socialized sector of the farms. There was certainly no shortage of reference to the dangers of using force and compulsion in order to bring about the desired sales, but such statements were accompanied by an urge to show rapid results. The outcome was predictable. As Wädekin phrases it, local officials 'knew quite well what was expected of them, and they acted accordingly.'[24]

In a move that was ominously reminiscent of Stalin's 1930 *Pravda* article against those who had become 'Dizzy with Success' and thus committed excesses during the campaign for mass collectivization,[25] Khrushchev, in February 1959, felt obliged to restrain those who had shown excessive zeal in their endeavors to trim the private sector. Less than a month later, an unsigned article in *Pravda* condemned 'violations of the principle of voluntariness in the purchase of livestock'.[26] The consequences of this change in the official stance, however, are probably best interpreted as a temporary disorder. With a full *carte blanche*, the campaign would surely have had even more drastic effects, but what actually did take place was no doubt quite sufficient in order to deal the private producers a severe blow.

Apart from direct administrative – and largely illegal – restrictions on plot sizes, the main thrust of the campaign was aimed at cutting the vital supply of feed for privately owned livestock. Since the plots proper are far too small to allow either grazing or growing feed for silage, the private sector is crucially dependent on help in this regard. As we shall see below, many of the conflicts between the two sectors revolve around precisely this issue. For the *kolkhozniki*, the *kolkhoz* charter provides loosely for such help from the *kolkhoz* (e.g. grazing rights and in kind payments of feedstuffs). This provides at least a token security. For the other categories of plots, however, not even this type of security exists. We shall have more to say in a moment about the formal rules in this respect.

The attack on feed supply was waged on two fronts. First, a special decree was issued to prohibit peasants from feeding their animals on bread and other food products bought in state and cooperative stores.[27] This seemingly peculiar step was aimed at curbing a practice that still looms large in Soviet agriculture, namely the quite rational act of selling produce to the state, at relatively high state procurement prices, and then purchasing bread in the state stores, at low state retail prices, to be used for feed.[28] The possibilities of efficiently enforcing such a law are obviously slim, as it is hard to tell from the look of a loaf of bread whether it will eventually be eaten by man or beast. It is also doubtful whether it had any real effects on the peasants. It is an indication of the magnitude of the problem, however, that the law is still in force. It has also been repeatedly underscored, most recently at a Politburo meeting in May 1985.[29] For those who have plots in urban areas, moreover, where no other feed sources are readily available, it can probably have more serious consequences. Most likely, it was also against this group that it was originally and primarily directed.

The second ingredient in the attack was aimed at restricting grazing rights and at curtailing the practice of making payments in kind for labour performed in the socialized sector of the farms. The latter in particular had long been (and still is) an important source of feedstuffs. During the upswing for the private sector in 1953–55, such payments had risen sharply in importance, and the gradual shift in policy during 1956 and 1957 caused only a slight decline to take place. After 1958, however, a sharp drop was registered. By 1960, payments in kind (in relation to labour input) were down by almost a quarter, compared to 1956–57.[30] Part of this squeeze must of course have been due to the fact that feed was desperately needed to build up herds in the socialized sector, and could thus not be diverted to the plots, but the general atmosphere of negative proclamations against the plots must surely have made it easier to discriminate (or harder to favour) the private sector. The consequences of the attack are clearly visible in Table 2.1.

Table 2.1 Indicators of private sector performance, 1956–64 (million hectares, million head, beginning of year)

	1956	*1957*	*1958*	*1959*	*1960*	*1961*	*1962*	*1963*	*1964*
Sown area	7.31	7.29	7.35	7.24	6.74	6.74	6.73	6.72	6.27
grains	1.66	1.50	1.52	1.39	1.21	1.18	1.16	1.16	1.02
potatoes	5.03	5.28	5.27	5.27	5.01	5.04	5.07	5.06	4.80
Cattle	27.3	28.4	29.2	29.2	25.0	23.0	23.9	24.5	24.1
Hogs	16.4	17.4	14.7	15.1	13.8	15.4	17.3	16.1	13.1
Sheep	20.1	22.9	25.8	28.6	28.8	28.1	29.6	29.9	26.6

Source: From Karcz (1966), p. 414.

The emergence of a 'support' agriculture

From the table we can see that total sown area in the private sector declined by close to eight per cent in the years 1956–60, almost all of which occurred in 1959–60. The bulk of the reduction was borne by grains, which were down by more than a quarter over the period as a whole. Potatoes, which had initially increased its sown area, declined by only some five per cent, a reduction in sown area which was limited to the years 1959–60.

This pattern provides a clear reflection of the importance of livestock production, with potatoes serving as feed, and from the table we can see that cattle holdings were not reduced until 1959–60. Two other typical patterns are also easily discernible: first that the hog population, which is easiest to rebuild, suffered the hardest, and secondly that sheep herds increased, as a substitute for larger livestock. Overall, we can safely conclude that the impact of Khrushchev's policy was substantial and that it must have had serious repercussions for urban food supply in particular.

Retreat and a Renewed Advance

Under the pressure that was created by the decline of the private sector, opposition to Khrushchev's discrimination against the plots grew rapidly. At the January 1961 Central Committee plenum, the campaign was called off. Its outcome had demonstrated the basic inability of the socialized sector to shoulder the responsibilities left by a decimated private sector, and when the May 1961 issue of *Kommunist* published a survey of various unmistakably critical reactions to the campaign, it was all over: 'Evidently the publication of this issue was a definite public signal for breaking off the campaign against the private sector.'[31] Khrushchev himself, however, did not see the writing on the wall. Two years later he would be back, only from a different quarter.

Much of what has been described above has concerned the *kolkhoz* sector, thus neglecting the state farms, the *sovkhozy*. At the time, however, the latter were growing rapidly, due chiefly to a policy of transforming weak *kolkhozy* into *sovkhozy*. In many respects, the difference between the two types of farms is not of any major consequence. When it comes to the private plots, however, there are some important ideological distinctions, which do have an important bearing on the discussion. The more precise nature of these will be dealt with in greater detail below, when the formal rules that surround the plots are discussed, but one part of that discussion is anticipated here, by noting that access to a private plot had never been considered a right for the *sovkhoznik*, in the way that it was for the *kolkhoznik*.

Whereas the latter was seen as a voluntary member of an agricultural collective, and thus a 'peasant' by nature, the former was considered to be a 'worker', and thus also ought to live like one. This attitude was reflected *inter alia* in the very Khrushchevian programme of building 'agrotowns' (*agrogoroda*), a programme which focused on promoting concentrated settlements of urban-type housing, and which in essence would mean an end to the keeping of private livestock, since there would simply not be anywhere to keep the animals.

The policy of converting *kolkhozy* into *sovkhozy* is important for our argument in that it meant – theoretically – that the share of 'workers' in the population was increasing, whereas that of 'peasants' was declining. In practice of course, such conversions would not automatically mean a removal of plot rights. On the contrary, it is expected that larger plots are found on *sovkhozy* that have been formed by conversion than on newly established ones. The point is entirely ideological, in the sense of legitimating a harsher policy against the private sector as a whole. Here was a golden opportunity to come back at that sector, and in so doing Khrushchev relied on the impeccable authority of Lenin:

> I would like to recall that V. I. Lenin, when he raised the question of *sovkhoz* organization, said that the worker should not have any livestock for his personal use. And that was absolutely right. We have deviated from this principle.[32]

In May 1963, the Presidium of the Supreme Soviet issued three new decrees, which in essence meant a renewed attack on privately owned livestock. The first of these underscored the prohibition against using produce bought in state stores for animal feed, threatening severe penalties, including confiscation of livestock and up to three years imprisonment for serious offenses. The second placed strict limits on non-taxable livestock holdings of the non-*kolkhoz* population, and the third placed prohibitive taxes on holdings above these norms.[33]

Again, the reaction was reduced supply and rising prices, and with the disastrous harvest of 1963 a general shortage of fodder developed in many parts of the country, for private as well as socialized animals. Forced slaughterings resulted, with the number of pigs in the socialized sector declining by as much as 32.8 per cent.[34] For the private sector, 1963–64 meant a renewed and even harder squeeze. From Table 2.1 above, it can be seen that total sown area declined by close to seven per cent, in addition to the earlier reduction, and that pig and sheep numbers also suffered considerably, the former by 18.2 and the latter by 11.3 per cent. With the parallel poor performance of the socialized sector, it is not hard to envisage what this squeeze on the plots must have meant to the crucial problem of urban food supply.

The emergence of a 'support' agriculture

The ousting of Khrushchev, in October 1964, meant a virtually complete return to *status quo ante*, as far as his agricultural policy was concerned. Symbolically, it even went so far as to reinstate in his previous position the former Minister of Agriculture Matskevich, who had been fired by Khrushchev. For the private plots it meant a lifting of all restrictions that had been imposed during the Khrushchev years. What was to follow for the private sector during the first part of Brezhnev's tenure was thus in some sense to be more of a retreat than of a long-term strategy. What the actual driving forces may have been behind private sector development during this period is difficult to say. The sharp downturn around 1970 was certainly striking, and one may be tempted to assume that the potential of development of that sector, within given rules and parameters, had by then been exhausted, leaving only regional and demographic factors to counteract its decline.[35] The present discussion, however, will dodge this thorny issue by skipping directly to 1977, when a conscious policy of encouragement and stimulation was introduced.

Lessons Learnt?

Before going on to examine that new policy, however, we shall conclude our digression into the Khrushchev era by listening to two different voices, commenting on two different aspects of the events just studied, beginning with Wädekin's summary:

> Thus Khrushchev's administration came to an end, without a change in the official attitude towards the private sector, but with increasing indications of a realization of the serious damage caused by his great campaign of 1958–1960.[36]

The lessons that can be drawn from Khrushchev's attempts at cutting the private sector down to size can be captured in a sentence that will be repeated a number of times throughout this study: 'Private is bad, but unfortunately it is also necessary.' As Wädekin indicates, the latter part of this statement has an unfortunate tendency to be neglected, with predictable consequences.

The second of our two voices belongs to Raig and will serve to illustrate that the periodic attacks that have been carried out against private activities may not necessarily be all malignant. They may – as in Khrushchev's case – rest on an honest conviction that curtailing the private sector is actually in the best interests of the country. The following vision of the future, as expressed by Raig, is typical of this view:

> The process of a gradual unification of socialized and personal agriculture will in the final analysis produce a situation where personal auxiliary agriculture loses its economic significance. Instead, its social functions will come to the fore, as it is largely transformed into a form of household activity. In this way, the remnants of traditional peasant organization and conditions of work will gradually disappear, and thus the differences in conditions of life between town and country will diminish.[37]

One would certainly be inclined to agree with this vision of how it *ought* to be. Yet, the pattern of development during the last decade has produced a move in the opposite direction, in the form of a pronounced revival of official interest in the private sector, and of a consequent growth in its economic importance. This revival of interest has proceeded in two qualitatively different stages. The first of these, which can be traced in three major decrees from the Central Committee, of 1977, 1981 and 1986, is marked by an emphasis on the important contributions made by the private sector, and by calls for increased support.[38] In contrast, the second stage is marked not only by calls for support, but also by important changes in restrictions on private sector activities. It can be traced in a decree from the Central Committee in 1987 and in the new 1988 model *kolkhoz* charter.[39] Before going into possible explanations and evaluations of this change in attitude, however, some of the more important characteristics of the private sector as it stands today will be briefly outlined.

The Private Sector of Today

A Changing Importance

Perhaps the most salient feature of the private sector of today is that the economic importance to households of undertaking various private activities has been substantially reduced. During many years of Stalin's reign, work on the plot was a necessity for survival, and in certain respects this situation can even be seen to have been a vital ingredient in the structure of the *kolkhoz*, a *sine qua non* for the system to work. The latter is described by Conquest in the following manner:

> The 'private plot' was a concession both to the peasant and to economic reality. But it was also an incentive to stay in and work for the *kolkhoz*. For it was to be taken away from any who did not put in the requisite number of 'labour-days' for the *kolkhoz*, and withdrawal from the *kolkhoz* naturally involved such forfeit. Thus underpaid labour on communal land was the condition of tenure – very much in the tradition of feudalism, in a stricter form.[40]

Today, of course, the system is different, with respect to both compulsion and the level of pay, and for most people work on the plot is a source of secondary income only. While the plots in 1940 accounted for 48.3 per cent of overall household income in *kolkhozy*, by 1965 that figure had dropped to 36.5 and in 1980 to 25.3 per cent.[41] Part of this fall has been due to higher payments from the *kolkhoz* but increases in various government transfers have also played an important role.

Given the long hours, and strenuous nature, of work on the plots, it can reasonably be expected that strong income effects will act to reduce labour input on the plot as public sector pay increases, and reality seems to confirm this expectation. In the 1959 census, 4.7 per cent of the population as a whole gave private plot activity as their main source of income (1.0 per cent for men and 7.8 per cent for women), while the figure in 1970 was down to 0.8 per cent (0.2 and 1.3) and in 1979 to merely 0.2 per cent (0.0 and 0.4).[42] A word of caution that is appropriate here, however, concerns the demographic aspects of private plot activities.

The importance of demographic patterns for production in the private sector may be illustrated by a sociological study, reported in *Literaturnaya Gazeta*, which shows a significant correlation between both age and education, on the one hand, and interest in animal husbandry (the traditional core of private output) on the other. For example, in families where one or both spouses had work requiring specialist competence (engineers, teachers, doctors, economists, agronomists, etc.) only 39 per cent held livestock. Furthermore, in families where both spouses were under the age of 30, only 34 per cent had cows or pigs. Perhaps most significantly, however, of all families with a resident *babushka*, the classic Russian grandmother, as many as 62 per cent were reported to have a cow.[43]

All of this illustrates that regional variations in demographic patterns will be a perhaps more important determinant of the performance of the private sector over time than will minor changes in official policy. Caution is thus called for in drawing conclusions from statistics that cover the union as a whole. Moreover, it also illustrates the likelihood that with rising educational levels and a change of generations, the plots may in the long run be transformed into a more 'normal' form of household gardening. As we shall see below, such a development is also what some Soviet sources envisage as 'normal'.

A second word of caution concerns the fact that there is a complicated interdependence between the private and the socialized sectors, which precludes simply looking at relative earnings possibilities. We shall deal with this at greater length in Chapter 4, but

Table 2.2 Major products of the private sector, 1970–84 (million tons; eggs, billion pieces; wool, thousand tons)

	1970	*1975*	*1977*	*1980*	*1981*	*1982*	*1983*	*1984*
Potatoes	62.9	52.3	49.3	42.0	44.7	49.3	49.8	n.a.
Vegetables	8.1	7.9	7.0	9.0	9.5	9.6	9.3	n.a.
Meat	4.3	4.7	4.3	4.7	4.6	4.6	4.7	4.6
Milk	29.8	27.9	27.6	27.1	25.5	24.4	24.1	23.3
Eggs	21.7	22.6	21.1	21.8	22.1	22.1	22.4	22.1
Wool	79.0	90.0	86.0	96.0	103.1	108.0	112.0	112.0

Source: Wädekin (1985b), pp.8–9.

we may note here that in spite of a rising level of pay from the socialized sector, such work may still be unattractive and the incentives not chiefly related to official pay. Michael Bradley, for example, quotes a Soviet source from 1965 as saying that 'members would not work on the collective at all in the absence of minimum work requirements.'[44] No doubt compulsion does play an important role, but there is also another reason to work for the socialized sector, a reason which stems from the important role that is played by animal husbandry in the private sector and from the consequent dependence on the socialized sector for feed.

Returning now to the main argument, we may note that the decline in economic importance of the plots to the households directly concerned is not by far mirrored by a corresponding decline in the importance of such activities to Soviet society as a whole. It is a fact that the *share* of the private sector in total agricultural output has been falling steadily, from 35.6 per cent in 1960, to 29.7 per cent in 1970 and 26.5 per cent in 1979, but this has been due largely to expanding socialized sector output.[45] After 1979, moreover, the relative decline has been halted.[46] If we look at the performance of the private sector in absolute terms, Table 2.2 shows that, with the exception of potatoes and milk, 1977 presents a turning point, after which output has been stable or even shown small increases.

Moreover, when we look at the private contribution by product we can see that it still remains of some considerable importance. This is reflected in Table 2.3.

From Tables 2.2 and 2.3, and from the figures given in the text, we can draw three important conclusions: (1) that the private sector accounts for a surprisingly high share of total agricultural output, (2) that its share in total marketed output is considerably lower, and (3) that its contribution is strongly concentrated to high value output (potatoes should to a large extent be seen as an input to the highly

Table 2.3 Private sector shares of output and marketing (per cent, 1985)

	Output	*Marketing*
Potatoes	60	41
Vegetables	29	14
Meat	28	13
Milk	29	2
Eggs	28	5
Wool	26	23

Source: Narkhoz (1986), pp. 185, 190.

profitable animal production). The latter two points are quite reasonable, given that the original intentions of allowing private plots were subsistence oriented, and that official Soviet price policy leaves room for specialization in highly lucrative products that are in great demand. The first point, however, may merit some comment. Let us begin by looking at the rules of the game.

The Rules of the Game

The right of Soviet citizens to have a private plot, with a private house and some private animals, is granted in very general terms by Article 13 of the current Soviet Constitution.[47] More specific rules on plot sizes and livestock numbers are laid down in special charters, applying to different social groups of the population. The main dividing line here serves to separate the *kolkhozniki* from the remainder of the population.[48] As the former are considered to be members of voluntary cooperatives, their rights are regulated by the *kolkhoz* charter, while the rules pertaining to other citizens are spelled out in republican legislation. This distinction illustrates the importance of ideological considerations, but it must also be recognized that the differences are largely the result of different historical origins. It is thus hardly surprising that they are clearly reflected in practice.

With the average size of a plot in the overall private sector, 'narrowly defined', being 0.24 hectares, a breakdown reveals that the largest plots will be found in *kolkhozy*, with 0.32 hectares, while those in *sovkhozy* register 0.21 hectares and those in other state enterprises and institutions amount to no more than 0.14 hectares.[49] Viewed in total terms as well, differences between the categories stand out, with 72 per cent of total plot land being found on the farms, and with a full 99 per cent of all *kolkhoz* households having a plot.[50]

As we have noted above, while the *kolkhozy* are considered to be agricultural cooperatives, with a voluntary membership, which form a separate and formally autonomous sector of the Soviet economy, the *sovkhozy* are treated in the same way as other state enterprises. One reflection of this separation is that while official decrees on agricultural policy normally give orders to the *sovkhozy*, the *kolkhozy* are only given 'recommendations'. As we have indicated above, this distinction has an important ideological bearing on the rights to private plots. While the *kolkhoznik* is considered to be a 'peasant', the category of 'workers and employees', including the *sovkhozniki*, is considered to be workers, like those employed in other sectors of the economy.

In the case of *kolkhozniki*, it must borne in mind that the rules on their rights to a plot were originally laid down in the 1930 and 1935 *kolkhoz* charters, and that we are thus dealing largely with a continuation of some of the main characteristics of the old peasant household. Although the 1935 charter was succeeded by the 1969 model charter,[51] differences between the two are only of minor importance.[52] It is not until the 1988 model charter that we can find important changes in the rules of the game.[53] The latter, however, will have a bearing only on speculations regarding the future course of events, so let us start by looking at such provisions that were made in the 1969 charter.

The formal conditions for being entitled to a plot are membership in the *kolkhoz* and fulfillment of established work quotas.[54] According to the charter, the size of the plot is normally limited to 0.5 hectares, but in reality there are wide variations. As stated above, the average plot size is less than two-thirds of the legal maximum. This is partly due to regional differences, where the presence of irrigation, for example, may cause reductions in the legal allotments, but past policy against the plots has certainly played its part as well. Allocation of land is formally decided by the *kolkhoz*, according to family size and participation in work in the socialized sector, but its general policy in such matters is no doubt subject to approval by local Party organs. No formal title deeds exist, nor can the *kolkhoznik* go to court in the case of a dispute.

Article 43 of the 1969 model charter lays down the following limits on livestock holdings: 'One cow with calves of up to one year, and one calf of up to two years, one sow with piglets of up to three months or two hogs for fattening, and up to ten sheep and/or goats.'[55] In addition, an unlimited quantity of poultry, rabbits and beehives are allowed. The basic principle of production on the plots has long been that no 'exploitation' may occur, which rules out renting land, hiring labour and leasing machinery, as well as possessing vehicles, major agricultural machinery and equipment, and productive livestock such as horses, bulls or oxen. Nor may the plot serve as a main source of income.

Having said all this, however, we must note that under the impact of Gorbachev's *perestroika* a number of traditional truths about the rules governing private plot production are now being actively challenged. The 1988 model charter removes centrally fixed limits on plot sizes and on private livestock holdings. Furthermore, it explicitly permits the possession of 'productive' livestock, and it provides for the allocation of additional strips of land to households willing to enter into various forms of contractual arrangements with the socialized sector of the farms.[56] Finally, and significantly, it also provides for allocating land to *kolkhozniki* in urban-type housing, which in essence means an end to Khrushchev's vision of the *agrogoroda*, the agrotowns.[57] Although it is obviously still too early to speculate about the potential impact of these changes, we may note that the removal of central restrictions does not imply a full *carte blanche*. These are now instead to be decided by the *kolkhoz* general assembly, which in essence means that it will be up to local Party officials. The 'reform' may thus in practice turn out to be quite conservative, in terms of granting local officials even more power than they already have.

The case of workers and employees is generally characterized by more severe restrictions.[58] In contrast to the *kolkhozniki*, however, there is no general charter to specify rules and regulations. As we have mentioned above, this is instead done in republican legislation, and here two different categories of citizens can be broadly identified. Those who live in rural areas and are directly or indirectly employed in agriculture, as agricultural workers or as specialists of various kinds (doctors, teachers, agronomists, etc.) have the most favourable position, whereas those who are employed by *sovkhozy* are of course particularly favoured. The size of plots in this category as a whole is considerably smaller than in the *kolkhoz* case. In *sovkhozy*, they should normally not exceed 0.3 hectares, but regional variations may extend up to 0.5 hectares. Particularly in cases where *sovkhozy* have been formed out of previous *kolkhozy* one may expect to find larger plot sizes. Those who are not employed in agriculture but live on the territory of a *kolkhoz* or a *sovkhoz* may be granted plots no larger than 0.15 hectares, in perpetuity or on a short-term lease.[59] Limits on livestock holdings are also more severe than in the *kolkhoz* case: one cow, one calf, one sow with piglets, and three sheep and/or goats are allowed, with some possibilities for substitutions.[60] In further contrast to the *kolkhoz* case, for this category as a whole formal title deeds are issued.

Before proceeding, we should mention in the context that there also exists a third category of plots, namely those belonging to workers and employees in urban areas. This category, however, belongs to the private 'fringe' and will be dealt with in Chapter 3.[61]

Let us turn now to look at the impression thus provided of this 'core' of the private sector.

Giant or Dwarf?

The picture of what is going on here is a bit difficult to interpret. On the surface, given the tiny size of the holdings and their substantial share of output, one would be inclined to agree with Shmelev in characterizing the private sector as 'giant and dwarf' (*velikan i karlik*).[62] This interpretation also seems to be shared by Lewin, for example: 'The private small enterprises look rather like a "giant dwarf", an important and vital producer.'[63] However, such a characteristic agrees rather poorly with reality.

The total amount of agricultural land officially at the disposal of the private sector, narrowly defined, stands at merely 8.11 million hectares. Of this, 6.15 million hectares is sown area, and the bulk of the remainder gardens and vineyards. If we place these figures in relation to the corresponding ones for the agricultural sector as a whole, it is found that the private sector has at its disposal a mere 1.33 per cent of all agricultural land, or 2.70 per cent of the total sown area.[64] Such figures would indeed lend support to the 'dwarf' impression, and it may be noted that these are also the ones most frequently cited, in both East and West. Shmelev, for example, normally gives 1.5 per cent, for agricultural land, whereas Alec Nove prefers 3.0 per cent, for sown area.[65]

If, however, we take into consideration the important role that is played by animal husbandry in the private sector, these figures cannot be accepted as credible. It is obvious from Table 2.4 that considerably more land must be available, for both silage and pasture. This, of course, is also the case. The problem, however, is that such use of public land is to some extent illegal, and can therefore not be advertised officially.[66] Instead, we have to rely on estimates of the true total area of land at the disposal of the private sector. One such is given by Ann Lane, at 119 million hectares.[67]

Table 2.4 Private sector livestock holdings (million head, January 1987; per cent)

	All Cattle	*Cows*	*Pigs*	*Sheep*	*Goats*
Total	122.1	42.4	79.5	142.4	6.5
Private	23.7	12.9	13.6	28.2	5.2
Private share	19.4	30.4	17.1	19.8	80.0

Source: Narkhoz (1987), p. 253.

If we assume all this 'extra' land to be pasture, we get a figure of 26.3 per cent as the private sector's share in total pasture, a figure which agrees rather well with its share in total livestock holdings. Without this breakdown, the figure for the sector's share in total agricultural land is 19.6 per cent, still a considerably more reasonable figure than the 1.5 or 3.0 per cent that were mentioned above. Even if we were to acknowledge that Lane's figure may be an exaggeration, including lands that are only very temporarily in use by the private sector, we are still left with a considerably different and more realistic picture than that with which we started.

To the uncertainty that follows from the lack of proper data on the direct utilization of land, ought to be added the unknown magnitude of such land that is indirectly placed at the disposal of the private sector, via in-kind payments of feedstuffs and via thefts from fields and barns. Such figures are of course difficult to produce (especially in the latter case), but all in all it is probably safe to conclude that considerably more resources go into private production than meets the official eye. Moreover, if we look at the total amount of labour that goes into plot activities, the distinction between 'dwarf' and 'giant' is yet further blurred.

In 1980, there were 32.5 million households that had at their disposal private plots of the 'proper' kind. Of these, 13.6 million were found on *kolkhozy*, 10.1 million on *sovkhozy* and the remainder with other state enterprises and institutions.[68] In 1983, the number of plots in *kolkhozy* had fallen to 12.6 million, while the figure for the latter two categories taken together had risen to 22.8 million, thus bringing the total up to 35.4 million. Moreover, if we should add in a further 11.2 million families found on the private 'fringe', which will be discussed below, we get a total of some 46.6 million families, which hardly seems to warrant speaking of a 'dwarf' sector.[69]

On the contrary, when calculated very roughly, with an average family size of 3.5, it is found that more than 160 million people, or well over half the Soviet population, have access to private plots. Here we find almost all *kolkhoz* households, about 80 per cent of workers and employees in rural areas and about 25 per cent of the urban households.[70] The socialized sector of Soviet agriculture has certainly come a long way from its original task of feeding the population.

One complication here is the highly important principle, which has been mentioned above, that private plot activity must not be a main source of income, i.e. that all able-bodied individuals must first carry out their work obligations to the socialized sector.[71] The latter is clearly reflected in the structure of plot labour input, the bulk of which is accounted for by people who for various reasons (children, mothers with infants, invalids, pensioners, etc.) do not take part in work in the socialized sector. According to a 1981 source, a full 83 per cent of all work performed in the private sector was carried out by such people.[72]

Figures from 1978, moreover, show that men in the labour active ages accounted for half the labour input in the socialized sector in *kolkhozy*, whereas their share in the private sector was only about 20 per cent. At the same time, the share of female pensioners in the socialized sector was less than 3 per cent, while their share in the private sector was almost 25 per cent. As Shmelev delicately puts it, the private sector 'widens the sphere of participation by the population in production.'[73] As we shall see below, in the Soviet literature this way of mobilizing a last labour reserve is considered an important 'social' function of the private sector. With this, let us conclude our discussion of the 'core' of the private sector.

Conclusion

The evidence that has been presented in this chapter has focused largely on dispelling some of those myths about the private plots that are frequent in the West – and encountered sporadically even in the Soviet Union. On the one hand, we have indeed presented a picture of important contributions being made, in support of the socialized sector, but on the other we have also shown some first signs of an argument which holds that the plots are very far from being a super-productive island in a decaying socialist agricultural sector. We have seen that considerably more land, labour and other inputs in various ways go into private production than meets the official eye.

An important additional ingredient in that endeavour has been to show some first signs of the pronounced ambivalence that marks the official attitude towards the private sector. That ambivalence has been illustrated via Khrushchev's attacks on the plots, which may well have been based largely on a sincere belief that the plots were set in an ideologically 'correct' phase of 'dying out'. In the final part of the chapter, however, we have argued that the plots of today are very far from that final phase, a fact which gives rise to considerable unease – ideologically and otherwise – with respect to their present as well as future roles. The fact that the plots show such stubborn resilience reflects more upon the poor performance of the socialized sector than on their own productivity. The latter is of course particularly so in the case of rural food supply.

All of what has been said so far has thus been intended to lead up to the conclusion that the really crucial question to ask, in this context, ought not be how it is possible for this 'dwarf' sector to make such a 'giant' contribution to the country's food supply, but rather how it is possible to allow the preservation within the large-scale, mechanized and socialized, official sector of Soviet agriculture of such a remnant of medieval agriculture, feeding on the labour of pensioners, children and invalids, and on the theft of other necessary inputs. Before proceeding to answer that question, however, we shall look briefly at the other two forms of auxiliary 'support' agriculture that were mentioned at the outset of the study.

Chapter three

The two fringes of Soviet agriculture

In addition to what we have referred to above as the 'proper' private plots, Soviet agriculture also has two interesting 'fringes', both of which have in recent years come to assume a growing importance. The first of these is made up of small subsidiary farms that are run by state enterprises and institutions, and even by units of the armed forces (including the navy!),[1] in order to provide food for canteens serving their employees. These are normally referred to as *podkhozy* (*podsobnye khozyaistva*), a term which will also be used in the presentation below.[2]

The second fringe is made up of small plots of land that are put at the disposal of urban residents who are not included in the 'narrow' definition used in the previous chapter, i.e. those who are not entitled to a plot via *kolkhoz* membership or via employment or other links to state agricultural enterprises. These plots serve the dual purpose of providing the urban population with both scarce produce for their tables, and with the possibility of a healthy and meaningful recreation, outside the congested urban areas. The latter in particular is currently receiving considerable attention.

There are obviously great differences between these two fringes, and at first glance it may even be argued that the former really has nothing at all in common with plot production, which is the topic of this book. They do, however, share one common trait, in that both perform an important support function for the official, socialized agricultural sector, and it is for this reason that the *podkhoz* sector has been included amongst the private plots. Furthermore, it should also be recognized that there are important differences between the 'proper' plots and those of the private fringe, differences which have called for our separation of the private sector into a 'core' and a 'fringe' part. To these differences, and to the rules that apply to the respective fringes, we shall return in more detail below. Let us start by looking at the official fringe – the *podkhoz* sector.

The Official Fringe

The history of the *podkhoz* goes back to the 1920s and the 1930s, but it was not until 1939, no doubt as a part of preparations for war, that it came to assume any degree of prominence. In the following year, a special Central Committee decree called for more such farms to be organized, partly on land from the state land fund, in order to provide food for factory canteens, and in 1942 another decree called for the 'further development' of this sector, which by then already numbered some tens of thousands of farms. During the difficult years 1942–44, with the exception of bread and flour, *podkhozy* supplied on average 10 per cent of the food intake of the urban population (in calorie terms) and in many cases enterprises were completely self-sufficient in food for their employees.[3] The importance that was officially ascribed to this new source of agricultural produce can be illustrated by the following excerpt from an *Izvestia* editorial:

> From an amateur occupation of a small group of persons in the recent past, it became, during the war years, a movement of millions – a potent source of improvement in the supply [of food] for the working people.[4]

By 1944, a total of about 16.5 million urban workers were engaged in this form of amateur gardening, a number which was 40 per cent higher than in 1943 and which exceeded its 1942 level by as much as two and a half times. At their disposal in the latter year, these new gardeners had 1.4 million hectares, or 83 per cent more land than in 1942. These rapid developments provide ample illustration of the emergency nature of the programme, and it is indicative of its perceived success that 1944 saw a further Central Committee decree, calling for an increase of approximately 20 per cent in the area allocated to such gardens, and for an increased provision of seeds, tools and crucially scarce mineral fertilizer.[5] The importance of this form of support, under conditions of emergency, can hardly be overestimated, nor is it surprising that the role of the *podkhoz* was greatly diminished in the post-war period.

What is more surprising, and hardly flattering for the development of the socialized agricultural sector, is that 1978 witnessed a further decree from the Central Committee, calling for a renewed support to the *podkhozy* in matters like the provision of seed and young animals, as well as strengthening of the so-called material–technical base, i.e. tractors, combines, mineral fertilizer, etc.[6] A similar – albeit more low-key – call for support was made in the 1982 Food Programme.[7] At the May 1982 Central Committee plenum, which launched that programme, Brezhnev stated that as a rule all state enterprises and institutions should have such farms.[8] As a result of this emphasis, great efforts were made to locate suitable land, and large sums of money were paid out in credits. Unfortunately, these efforts do not seem to have had quite the kind of results that might have been expected, but before going on to examine their record the ambitions will be briefly examined.

Goals of the Podkhoz Policy

Two important goals can be discerned behind the process of strengthening the *podkhoz* sector. First, there is the fairly obvious one of welcoming any addition to the country's total food supply, a goal which should be seen against the previous background description of the performance of the 'real' farms. It is evident from the decrees referred to above that such support is critically needed. It may also be illustrated by the fact that members of the families of those working in the enterprises or institutions in question often participate in the work on such farms.[9] Much of our evaluation of the performance of this new sector will thus focus on the extent to which it has been possible actually to mobilize – and utilize – some idle last reserves. To that we shall return in a moment.

The second goal is perhaps somewhat less obvious, but definitely no less important. First of all, it seems clear that we are not simply dealing here with an ambition for factories and units of the armed forces to produce food for their own needs. In addition, there is also a growing desire to take some general strain off the official system of procurement and distribution of foodstuffs. As we have mentioned above, the performance of this system leaves quite a lot to be desired, particularly with respect to distant regions, with poor transport and unfavourable conditions for local agriculture. There is a clear policy in such cases to place an increased responsibility on enterprises to cater for their own needs, by developing their own *podkhozy*:

> Experience shows that in the first stages of industrial development of new regions it is particularly effective to create *podkhozy*, specializing in various forms of perishables and foodstuffs that are difficult to transport: milk, meat, fowl, eggs, potatoes and vegetables.[10]

This policy of attempting to cope with difficult supply problems by shifting responsibility onto the shoulders of the employers of those who are to be fed is referred to below as 'structural autarchy', and it is far from specific to industrial enterprises in newly developed areas.

A similar objective can be identified in the case of popular resort areas, where a severe pressure is placed on the official supply system during a brief season of peak demand. In the Krasnodar *krai*, for example, which is situated on the attractive Black Sea coast and has a resident population of only 5 million, there are annually some 12 million 'official' holiday makers in sanatoria and hotels, and an additional unknown number of 'wild' visitors (*dikary*), who add their pressure of demand on local markets and food stores. An indication of the share of the latter is given from the Baltic resort of Yurmala, where they are reported to constitute no less than 20 per cent of the official ones. In order to ensure a reliable supply of foodstuffs for their hotels, the labour unions, which are responsible for a large proportion of the official network of hotels and sanatoria, have developed their own *podkhoz* system, which in 1980 numbered some 400 farms. Together, these supplied about 10 per cent of the food requirements of the 10 million or so that got their vacations via the Unions.[11]

Another, and perhaps somewhat more peculiar indication of an ambition for structural autarchy can be found with respect to the armed forces. For obvious reasons, much secrecy surrounds this sector,[12] but in simple terms, we are dealing with three types of farms. First of all, we have the military farms (*voennye sovkhozy*) in the supply system of the Ministry of Defence. Second, we have *podkhozy* attached to military enterprises and organizations, and finally there are small 'kitchen enterprises' (*prikukhonnye khozyaistva*), which are attached to the actual military units.

One indication of the growing importance that is being placed on military food production was given in 1982, by the then head of Army logistics, General Kurkotkin, who presented the role of such farms as particularly important in remote areas, and who also placed special emphasis on the role of the 'kitchen enterprises', in providing food for the tables of soldiers and sailors.[13]

In another statement Admiral Mikhailovsky, then commander of the Northern Fleet, referred to the outstanding performance of a model navy *sovkhoz*, the *Severomorets*, as a demonstration to the sceptics of the potentials for successful agriculture north of the Arctic Circle. In the same article he also claimed that all the milk and eggs consumed by the Soviet navy were provided out of its own sources.[14]

Following the adoption of the 1982 Food Programme, plans for military farms were hurriedly and drastically increased, raising plan targets for that same year by some 15 per cent and envisaging a 50 per cent increase in output by the end of the 1981–85 plan period, as compared to the annual average of the preceding five-year period.[15] Of particular interest here is the great stress that was placed by General Isaenko, then head of the Ministry of Defence victualling service, on achieving increases primarily in the Far East, the Far North, the Transbaikal, and Central Asia, all in order to reduce pressure on the normal supply lines from the southern and central parts of the country.

Of great interest is also the General's call for support to private plot farming of the families of servicemen.[16]

All of these statements are of obvious relevance for our discussion of structural autarchy, as they bring out the shortcomings of the 'regular' farm enterprises in meeting their classic obligations of providing food for the cities and the army. Supplying land, animals, tools and implements, and even greenhouses for operational units of the armed forces would hardly seem to be part of the traditional assignments of the miltary logistics service.

Finally, the consumer cooperation Tsentrosoyuz, which is charged with taking produce from field to consumer, is also busy developing its own system of *podkhozy*. Traditionally, it has focused chiefly on developing feed lots for the fattening of cattle. Thus, in 1981 there were some 8,000 such farms, having together 586,000 hectares of agricultural land, out of which only 70,000 was sown area. The latter illustrates the strong emphasis on animal production, and livestock holdings amounted to no less than 70,000 head of cattle, 370,000 pigs and 200,000 sheep and goats. Total output was given at 81,000 tons of meat and 12,700 tons of fish. Seen in the overall context, these figures are of course not very impressive, but for a *trading* organization they have a quite important message to tell. Recently, moreover, a policy seems to be unfolding which focuses on forming highly specialized *podkhozy*, with the aim of integrating the full chain of production, from procurement of feedstuffs to slaughter and meat processing.[17] All of this clearly indicates a desire by the Tsentrosoyuz to move away from its original intended function – that of procuring from the farms – to become itself a producing enterprise, with its own outlets.

In this first step of the presentation of *podkhoz* policy, it can be seen how an emphasis on structural autarchy has aimed at a reduction of the dependence between various branches of the economy, thus sacrificing economies of scale in specialization, in order to relieve some of the pressure on the official supply system. This ambition, however, does not cover the full picture regarding the role of the *podkhoz*, and we shall thus proceed now to see how additional burdens have come to be placed on these stand-in farms.

A New Role for the Podkhoz

Initially, the role of the *podkhozy* was surely no more than one of self-sufficiency, with the 'parent' enterprise having full responsibility for its own farming operations. This policy, however, did not fare particularly well. At best, enterprise management seems to have accepted the new responsibilities grudgingly, and probably for good reasons. First, it has been a well-known practice for quite some time for local authorities in trouble to issue calls for industrial enterprises in their respective areas to 'help out' with the harvest, in any way possible – manpower, transport, machinery, repairs etc. This issue will be discussed at greater length below, but for the moment we may note that in order to hedge against such disruptions in production which must follow from these emergency calls, enterprises are forced to engage in a well-known and no doubt costly overstaffing, which in essence amounts to an obligation to carry on their payrolls a large section of the seasonal agricultural labour force. The evidence, moreover, of what takes place when this 'help' arrives on the farms is at best anecdotal, and it would seem difficult indeed for any of the involved to be satisfied with this state of affairs.

In addition to this forced 'parenthood' towards failing farms in their respective areas, the 1978 decree on strengthening the *podkhozy* also called upon Soviet state enterprises and institutions to accept an increased responsibility for farming operations of their own, an endeavour for which most have surely been ill prepared and for which only minor help (if any) has been awarded. Land of adequate quality has been difficult to find and the notorious shortages of machinery, spare parts and fertilizer that have long plagued the 'real' farms, have hardly been eliminated. Perhaps most seriously, however, interest in the development of *podkhozy* has been stifled by the emergence of a form of quite illegal procurement practices.

One Soviet observer of *podkhoz* development flatly notes that in some cases 'no sooner has an agricultural workshop taken its first steps, still being held under the arms, than it is ordered to deliver half of its output into the state fund.'[18] In another case, we are told how interest in developing these farms is weakened by 'the persistent and illegal practice of planning the delivery of *podkhoz* products as contributions towards the state procurement targets.'[19] Maybe it was against this background that the authorities decided to go all the way, by incorporating the *podkhozy* fully into the overall system of state procurements. The signals in this direction that were contained in the 1978 decree on support for such farms have been described in the following way by *Pravda*: 'A very timely document, making it possible now to carry out these useful activities not in an independent fashion, as it used to be, but in a planned, organized manner.'[20]

Two steps taken in 1979–80 both indicate the latter desire. As of January 1979, it was decided that deliveries of meat and milk from these farms to the canteens of their 'parent' enterprises should take place at the fixed state procurement prices, and not as previously at the considerably lower retail consumer prices (the 'purchasing' enterprise would be compensated for the increase in costs). The aim of this step was clearly to strengthen the economy of the farm sections.

In a second step, in 1980, a system was devised for state purchases to take place out of surpluses not used by the 'owning' enterprises.[21]

The picture that transpires from this presentation of *podkhoz* policy is thus one of a rather anomalous coexistence of two trends of development, both of which are unsatisfactory in their own rights. On the one hand, there is the ever more serious labour constraint on the 'regular' farms, i.e. the *kolkhozy* and the *sovkhozy*, which forces industrial enterprises to engage in costly salvage operations, and on the other, the recent ambition to have these very same enterprises make up for the poor performance of their 'adopted' farms, by mounting farming operations of their own. It should not take much thought to realize that most enterprises are ill equipped for both of these tasks, and that once they find themselves trapped in this situation, it is logical for them to demand the right to merge the two operations, in order at least to realize some of the potential gains from specialization.

In a round table discussion on the problems of 'backward' farms, organized by *Izvestia* in 1981, it was quite frankly suggested that *kolkhozy and sovkhozy* that were 'chronically' in need of help should be placed under the direct control of industrial enterprises in their area.[22] This rather drastic suggestion was made, for example, by an industrial manager who claimed that if he were allowed to take over a nearby *sovkhoz*, he would not only promote an economic revival of the village, but also succeed in raising output. Meat production would increase by some 100–150 per cent, and the overall effect would be not only that of securing a stable supply of foodstuffs for his factory canteens, but also an improved general supply of vegetables and potatoes for workers.

The suggestion was strongly seconded by *Izvestia's* correspondent, and on the following day the paper carried a full page of materials on these 'green workshops' (*zelenye tsekhy*) of industrial enterprises.[23] Here a chief engineer at a heavy machinery works says, in an interview, that if his enterprise were allowed to take over – as 'workshops' – a couple of nearby backward farms, it would not only turn these into profitable enterprises but would also assume in full their state delivery obligations. Any produce in excess of these obligations would be used in the factory canteens. Again, *Izvestia* supports the suggestion and argues that it is necessary not only to arrange for a transfer of disused land to industrial enterprises and state institutions, but in some cases also to place under their direct control neighbouring backward *kolkhozy* and *sovkhozy*. Interestingly, in a major 1987 decree on *podkhoz* agriculture precisely such measures are suggested, 'if necessary', as a way of solving the problems of weak and unprofitable farms.[24]

The two fringes of Soviet agriculture

The picture of what is going on here is dominated by two separate strands of development, each of which is peculiar in its own right. On the one hand, we have the development of 'green workshops' in industry, described above, and on the other we have the increasing needs of the farms to produce their own make-shift implements and spare parts, which is due to the poor performance of those responsible for 'material and technical supply'. In our discussion below, of the private fringe, we shall see how a similar 'reversal' of responsibilities is taking place between the urban population and the 'peasants', as the former is increasingly turning to the land while the latter is turning away from it. The absurdity of this situation has been captured rather vividly by Andreas Tenson, in his speculation that it 'may thus be that the golden age of communism triumphant will witness idyllic scenes of industrial workers milking cows and bucolic peasants fashioning nuts and bolts.'[25]

Performance of the Podkhoz Sector

If we turn now to look at the performance of the *podkhoz* sector, a rather bleak picture presents itself. First of all, it is doubtful whether the ambitions for regional and structural autarchy have been realized. Available information indicates a fairly uneven development on this count. For example, while *podkhozy* in the RSFSR in 1980 supplied on average 1.27 kg of meat *per capita* of the urban population, more than 3 kg was noted for Rostov *oblast* and close to 4 kg for Omsk *oblast.* Similar differences can be found between ministries, with the Ministry of the Coal Industry having an average of 1,124 head of cattle per farm on its *podkhozy*, and the Ministry of Non-ferrous Metallurgy registering merely 136 head. At the beginning of 1981, there was an overall total of 12,900 *podkhozy*, while the number of enterprises and institutions that ought to have such farms was reported to be 2.5 times higher.[26] As Shmelev puts it, in the 'work of the *podkhozy*, there remains a considerable number of unresolved issues.'[27]

Although its development is uneven, if we look simply at the *number* of farms, we get a definite impression of a rapid development of the *podkhoz* sector, particularly so in recent years. Following the 1978 decree, there was a spurt of activity, causing the number of *podkhozy* to increase by no less than 2.5 times in the years 1978–83, thus bringing their total up to 21,200. By then, the sector comprised some 3.8 million hectares of agricultural land, of which 1.4 million was sown. It also had a total of 621,900 head of cattle (of which 183,600 cows), together with 2,714,900 pigs and 555,100 sheep and goats.[28] For the sake of comparison, we may mention that there were in 1983 a total of merely 26,000 *kolkhozy* and 22,313 *sovkhozy*.[29] In these terms, the growth of the *podkhoz* sector undoubtedly seems rather impressive. A study of the output figures reinforces this impression. According to one Soviet source, *podkhoz* production during 1978–81 increased by as much as 16 per cent for meat, 40 per cent for milk, 10 per cent for eggs and 24 per cent for potatoes.[30]

It is perhaps most important to note in this context that the official pressure for growth of the *podkhoz* sector has also been maintained in the period after 1983, i.e. after the attention surrounding the introduction of the 1982 Food Programme had started to fade. According to figures in the 1987 edition of *Narodnoe khozyaistvo*, by 1986 the number of *podkhozy* stood at 20,900. Together these farms had at their disposal a total of 5.2 million hectares of agricultural land (of which one-third was sown), and 908,000 head of cattle (including 263,000 cows). Compared to 1983, the absolute number of farms has thus been reduced somewhat, whereas the average farm size has been substantially increased. Total land is up by 36.8 per cent and the total number of cattle by 46.0 per cent.[31]

While it should be obvious that a comparison with the large-scale *kolkhozy* and *sovkhozy* in terms of area and livestock holdings is rather pointless, it may be of a somewhat greater relevance to make such a comparison with the private plots. The *podkhoz* sector of today has at its disposal almost two-thirds as much agricultural land, but very considerably smaller livestock holdings, corresponding to merely 3.8 per cent of the cattle and 18.6 per cent of the pigs in private possession.[32] This seeming imbalance is of course explained by the fact that a large amount of land is unofficially at the disposal of the private sector, for pasture and silage. The general impression of *podkhoz* development in this perspective is thus clearly that of a fringe, albeit rapidly growing, subsector of Soviet agriculture.

The impression of a fringe is further reinforced when a comparison is made with figures given above on the share of output for the 'proper' plots. In spite of its rapid growth, by 1982 the *podkhoz* sector accounted for only 3.1 per cent of all meat, 0.7 per cent of the milk, 1.0 per cent of the eggs, 1.8 per cent of the vegetables and 1.4 per cent of the potatoes produced in the socialized sphere (i.e. exclusive of the substantial private plot output).[33] Figures given by Ioffe for the same year indicate shares of only 1.8 per cent for meat, of 0.5 per cent for milk, and of 1.0 per cent for eggs.[34] Presumably, the latter apply to overall production, i.e. including the plots. By 1986, *podkhozy* still produced no more than 2.0 per cent of the total output of meat, 0.6 per cent of the milk, 0.6 per cent of the potatoes and 1.5 per cent of the vegetables that were produced by Soviet agriculture in that year.[35]

An obvious conclusion that can be reached regarding the role of the *podkhozy* is that they serve, in traditional Soviet fashion, to mobilize a 'last' reserve. Viewed in terms of the sheer *number* of farms, the impression of success is fairly evident. At the same time, it should be equally obvious that this in no way presents a solution to the problems facing Soviet agriculture, particularly in terms of the opportunity costs incurred, i.e. the output forgone when industrial resources are diverted into subsidiary agricultural operations. As we shall argue in more detail below, it may actually be the case that in a long run perspective more harm than good is done. A similar conclusion will be reached when we now turn to examine the second of the two fringes.

The Private Fringe

The private fringe of Soviet agriculture, i.e. such plots of land that are made available to urban residents who are not members of *kolkhozy* or employed by *sovkhozy* or other state enterprises or institutions, is at the same time the most interesting and the most controversial aspect of the private sector broadly defined. The official terminology is unfortunately a bit vague here. First of all, there are really two separate parts – one organized and one unorganized. The latter represents individual houseowners, in urban as well as in rural areas, with gardens in productive use. For obvious reasons, data on this part are difficult indeed to come by, and our main interest shall thus be directed at the organized part of the fringe, which does figure in official publications.

Most frequently, the latter is broadly referred to as 'collective household gardening' (*kollektivnoe sadovodstvo i ogorodnichestvo*), while when it comes to its actual organizational form, we normally encounter the term 'gardening associations' (*sadovodcheskie tovarishchestva*). The main problem here lies in the fact that there are several different kinds of associations, pursuing somewhat different activities under somewhat different rules. When they are encountered in statistics or in newspaper or journal articles, it is not always clear to which of them reference is being made. It may thus perhaps be justified to start by trying to establish the main terminological points.

The most important distinction is that between *sadovodstvo* and *ogorodnichestvo*, a distinction which is somewhat difficult to render in English.[36] Technically, there are two possible principles of classification. First, *sadovodstvo* strictly refers to the growing of fruits, berries and grapevines, whereas *ogorodnichestvo* denotes vegetable, potato and melon growing. *Sadovodcheskie tovarishchestva* might thus be referred to as 'orchard associations', whereas *ogorodnicheskie tovarishchestva* could be termed as 'gardening associations'. Such a simple distinction, however, is confused by the fact that there seems to be a growing practice of allowing the 'orchard associations' to engage in 'gardening' as well, thus making *sadovodstvo* a catch-all term for a great variety of activities:

> In order to improve the efficiency of land use in collective orchards, the model charter provides for the growing of not only fruit and berries but also vegetables. Such a practice does not imply transforming the orchard into a garden, but merely a greater diversity of production, a more varied supply of fruit, berries and vegetables in the household diet.[37]

The other technical distinction refers to the period of tenure. While charters for *sadovodstvo* are normally granted for longer periods of time, *ogorodnichestvo* is chiefly intended to be temporary, perhaps to last only for one season. The reason for this distinction is linked with the source from which land is allocated. If suitable strips of land are available in locations where no alternative use is planned for the forseeable future, they may be profitably used for the formation of a more permanent 'orchard association', whereas if we are dealing with plots of land that for various reasons may be only temporarily avail- able, a more short-lived 'gardening association' would be the logical choice. In consequence, there are firm rules against planting trees and bushes, and against erecting any form of buildings, on the latter. Again, however, even this seemingly clear-cut distinction is plagued by fluid border lines. The latter is brought out in the following way by a Soviet source:

> Unfortunately, as a result of poor monitoring by local organs, in practice these rules are sometimes broken, as fruit trees are planted and small houses built. This means that where in the coming 5–10 years the plan may have envisaged forest plantations, water reservoirs, public nature reserves, etc, a quite unplanned orchard will be growing up. Confiscation of land, which in such cases legally and inevitably must occur, will often be accompanied by complaints, conflicts, etc.[38]

The main reason for confusion on this point is historical. The rights of workers and employees to receive small plots of land for individual vegetable gardening purposes dates back to a decree of 1933, and is thus intimately connected to the rights of *kolkhozniki* to have private plots.[39] During World War II, this form of urban gardening assumed a significant importance in overall Soviet food supply, and a 1942 decree called for more attention to be given to such activities.[40] Here the parallel goes to the simultaneous expansion of factory farming, the *podkhozy*. As in the latter case, the importance of individual gardening declined after the war. The post-war years, moreover, would also see an important transformation in the nature of these activities, which thus far had been mainly individual and mainly, if not exclusively, focused on vegetable growing.

A disturbing feature was that the more permanent forms of gardening associated with *sadovodstvo* were gaining in importance. This obviously called for the issue of individual gardening as a whole to receive a more systematic attention, i.e. for detailed regulation to be introduced. Thus, in 1961 a decree was issued which called for a crack-down on this spread of *de facto* private property rights in the land.[41] In a parallel process, legal provisions were made for the formation of collective associations for fruit and berry growing, the *sadovodcheskie tovarishchestva*. The development of the latter shall receive more attention in a moment. An interesting point to note here, however, is that formal legislation was extended to cover only *sadovodstvo*. A 1984 Soviet legal handbook, for example, explicitly acknowledges that the rules governing *ogorodnichestvo* are not firm, being based chiefly on contractual arrangements.[42] Similarly, the Soviet agricultural encyclopedia recognizes that while *ogorodnichestvo* can be either individual or collective, *sadovodstvo* is strictly collective.[43] In contrast to *sadovodstvo*, property rights for *ogorodnichestvo* are consequently not vested in the gardening collective, but rather in those enterprises and institutions that employ the 'gardeners'.[44]

It is tempting to conclude from the above that the distinction between *sadovodstvo* and *ogorodnichestvo* is in practice not always quite clear, even to Soviet authors on the subject. This suggests that any data presented should be used with perhaps even greater care than usual. In addition, we might also mention that there are a number of other similar associations which add their share to the burden of confusion, such as common *kolkhoz* or *sovkhoz* gardens (*obshchestvennye ogorody*), cattle-rearing associations (*kollektivnoe zhivotnovodstvo*), rabbit-breeding associations (*kollektivnoe krolikovodstvo*), and maybe others as well, the inner mysteries of which escape the present author. These latter forms, however, shall be largely ignored in our further presentation.[45]

Given this terminological confusion in the Soviet sources, and for conformity's sake, we shall have to adopt, somewhat reluctantly, a rather sloppy terminology below. Thus, when referring to the organized part of the private fringe quite broadly, we shall (erroneously) use the term 'gardening associations', whereas when reference is made to specific sources we shall attempt to exercise a somewhat more careful approach. Before leaving the problems of terminology altogether, however, there is one final important point to be raised, which concerns the strong emphasis that is placed on the 'collective' and organized nature of the activities in question. From an ideal Soviet perspective, we are (in theory) not dealing with spontaneous private activities, but rather with controlled associations which are incorporated into the overall system via formal laws and/or officially approved charters that regulate their activities. This point will play an important role in the subsequent discussion.

A Rapid Growth

It is difficult to present a coherent picture of the development of the private fringe, even more so perhaps than in the case of the official one. Partly this is because only scattered data are available, partly it is due to the special definitional problems outlined above, and partly it is because reference to the number of participants is sometimes made in terms of families and sometimes in terms of individuals. In addition, some sources refer to the private sector broadly defined, i.e. including both the fringe and the proper plots, without indicating a breakdown for the different categories. Raig, for example, claims that there was in 1983 a grand total of 46.6 million families involved, who together cultivated some 8.5 million hectares.[46] The following is an attempt at piecing together whatever evidence is available.

In 1951, there were about 40,000 people who, joined together in orchard (*sadovodcheskie*) associations, had at their disposal some 3,400 hectares of land for private gardening purposes. By 1982, membership exceeded four million individuals, on a land area of close to a quarter of a million hectares. Some 70 per cent of these by now fertile lands had once been ravines, marshes and other less useful lands.[47] As in the case of the *podkhoz* we can thus note a policy of mobilizing some last reserves, this time in the form of disused land and the labour of pensioners and vacationers.

In comparison with the recently rapid development of the *podkhoz* sector, the growth of the private fringe presents a perhaps even more explosive impression. During the years 1980–82 alone there was – according to Shmelev – an increase in the number of participating families by no less than 1.3 million, or by more than half the total growth of 2.3 million for the entire 1951–82 period.[48] In a subsequent publication by the same author an increase by 46.1 per cent, or by 3.5 million families, is reported for the years 1977–83.[49] The seeming discrepancy derives from the fact that the former figures refer to *sadovodstvo* alone whereas the latter covers *ogorodnichestvo* as well.

Data for single years offer the same problems. In a 1981 publication, for example, Kubyak and Maksakova claim that more than 3.5 million families of workers and employees, in 24,000 orchard (*sadovodcheskie*) associations, together cultivated some 220,000 hectares of land. About 900,000 of these families had received their plots during 1978 and 1979.[50] In 1982, *Izvestia* mentions a figure of 15 million workers and employees, including family members, as having access to orchard plots (*sady*),[51] while in a 1983 publication, Shmelev notes that close to 10 million people are 'presently' involved, and that their numbers are growing rapidly.[52] Two years later, the same author is more specific, claiming that in 1983 a total of 11.2 million families were involved.[53]

The seemingly wide divergence between these various sources is again explained by definitional problems. In a 1983 publication, sociologists Karakhanova and Patrushev report more than 3.5 million members of orchard (*sadovodcheskie*) associations which, including family members, corresponded to 11 million people, or about 6.5 per cent of the urban population. In this category, there was some 240,000 hectares of land under orchard (*sadovodcheskie*) plots. At the same time, about 20 million people, or close to 12 per cent of the urban population, were reported to have access to garden (*ogorodnicheskie*) plots, on a total of 350,000 hectares. In the country as a whole, one out of fifteen urban families had access to an orchard plot (*sad*).[54]

The latter figure on membership seems to be a slip-up, due to a confusion of definitions. The respective figures for land also seem difficult to compare with Shmelev's data for 1983, which indicate a total of 752,000 hectares for both categories (up by 37.2 per cent since 1977). The latter also provides a figure of 29,400 for the total of orchard (*sadovodcheskie*) associations at the beginning of 1984 (up 7 per cent from 1983), which largely concurs with an assumed increase from Kubyak and Maksakova's earlier figure of 24,000. Shmelev is also one of the few to provide information on membership size, stating that 'some' associations may have 500–600 members, whereas 'others' may have over 5,000.[55]

In 1985, Gorbachev claimed that 20 million people were involved,[56] and in the 1986 edition of *Narodnoe khozyaistvo* a figure of 5.0 million families, on 179,000 hectares of land, was given for the former (*sadovodcheskie*), and of 5.6 million families, on 409,000 hectares for the latter (*ogorodnicheskie*).[57] As a final illustration of confusion, we may note that information in the 1987 edition of *SSSR v tsifrakh* indicated 6.3 million families on 402,300 hectares for the orchards (*sady*), and 5.6 million families on 408,500 hectares for the gardens (*ogorody*).[58] Both sources refer to 1985. In the first case, the total comes to 10.6 million families, while the second adds to 11.9 million families, neither of which is anywhere near the Gorbachev figure of 20 million people. The wide divergence in data on land in the orchard associations may be partly explained by the fact that the former source refers to cultivated land only. Neither indicates the respective numbers of associations.

Table 3.1 Development of gardening and orchard associations (thousand families, thousand hectares)

	Gardening Associations		*Orchard Associations*	
	Families	*Hectares*	*Families*	*Hectares*
1970	4,935.1	362.0	2,327.6	154.2
1980	4,914.0	364.5	4,084.6	257.4
1985	5,622.0	408.5	6,280.4	402.3
1986	5,697.9	414.2	6,852.6	439.9

Source: Narkhoz (1987), p. 237.

With the appearance of the 1987 edition of *Narodnoe khozyaistvo*, finally, a semblance of statistical order and consistency is introduced. Table 3.1 summarizes development over the past decade and a half, as it appears in this source. One immediate impression from this table and from the figures presented in the previous text, is that no two sources – not even the 1986 and the 1987 editions of *Narodnoe khozyaistvo* – seem to present a mutually consistent picture. Another striking feature is that it is the orchard associations that have been responsible for the bulk of the growth, with an increase of 194 per cent in the number of families and of 185 per cent in land area during 1970–86, whereas gardens have grown by merely 15 and 14 per cent respectively. This development illustrates clearly that the private fringe is taking on a more permanent nature, as the share of orchards is growing. Maybe this can also explain some part of the confusion found in the previously cited sources.

Putting this simple numbers excercise on one side,[59] however, a more important fact to be noted here is that growth has indeed been very rapid. This is further underlined by the fact that demand still seems to be very far from satisfied. In the 1986 decree on support for the gardening associations, for example, an additional growth of 1–1.2 million orchard (*sadovodcheskie*) plots annually was envisaged for 1986–90.[60] With a total of 11.9 million families involved in 1985, 12.5 million in 1986, and a further addition of maybe 4–5 million in the coming years, it may well be that by the end of the decade some 50 million members of the Soviet urban population will be involved in plot agriculture, in addition to 20 or 30 thousand *podkhozy* and the 35 million or so of 'proper' plots. How will such an agricultural sector be characterized? We shall have reason to return to this important question in the concluding chapter of the study.

In order to understand the controversial nature of the private fringe, it must be realized that we are dealing here with two very different categories of plot. On the one hand are members of gardening associations and urban houseowners with gardens (*usadebnye uchastki*), and on the other inheritors of rural plots together with those having bought or wishing to buy a house – a *dacha* – in rural areas.[61] The first of these two categories represents an 'ideal' picture. Such plots have long since been fairly well accepted, and for the gardening associations there has even been a number of model charters laid down, to control and regulate their activities.[62] As we shall see below, it is around the latter that the more heated part of the debate turns, but before going into this area let us look at the more 'acceptable' part of the fringe.

The Gardening Associations

Gardening associations are as a rule organized in connection with state enterprises and institutions, for the use of their employees. According to the Soviet agricultural encyclopedia, the basic function of such associations is 'to organize collective gardens [*sady*] in order to meet the needs of workers and employees for fruit and berries, as well as to provide a possibility for recreation and for educating children into sound working habits.'[63] In urban areas, land for such purposes is allocated to the enterprises and institutions in question by the city Soviets, while in non-urban areas, land is normally taken from the state land or forestry funds. In some cases it may even come from *kolkhozy* or *sovkhozy* with vacant land.[64] The decision to accept new members is taken jointly by management and the local labour union committee.[65] In so doing, particular emphasis is placed on the quality of work performed by the potential new member, i.e. the right to a plot should be part of the overall system of 'material stimulation'. *Trud*, the central daily of the labour unions, expresses this principle in the following manner: 'A correct behaviour is shown by those union committees which first of all accept as members cadre workers, veterans, shock workers and outstanding participants in socialist competition.'[66]

Official restrictions on this form of collective gardening have become considerably more tolerant over time. In the case of the RSFSR, a first model charter was worked out in 1956. This in essence permitted only the cultivation of land.[67] Then, in 1966, a revised version was accepted, which allowed both the keeping of animals and the building of small houses.[68] Following the attention that was given to these forms of amateur gardening in the 1977 decree on further development of the private sector, yet another revision was made in 1978, laying down in quite some detail the revised rules of the game.[69] According to the latter text, association members may build a small house for family use during the summer, with a living space of 12–25 m^2 and with certain heating arrangements. There may also be a terrace of up to 10 m^2, and if a basement cannot be fitted under the house a special cellar of 6–8 m^2 may be constructed, in excess of the above limits. Furthermore, a shed of up to 15 m^2 may be constructed, for the keeping of rabbits and fowl, as well as tools and implements. Livestock holdings may not exceed 20 head of fowl, 5 rabbit does and 5 beehives. The size of the plot, finally, may not exceed 600 m^2, 60–65 per cent of which should be cultivated.[70] If located outside urban areas, however, the plot may cover twice the given size.[71]

While this mania for regulation can certainly be said to be typical of the Brezhnev era, it is equally typical of Gorbachev's *perestroika* that a decree of September 1987 would remove much of the previous detail. Here it is stated in broad terms that members have a right to build houses, storage sheds and even greenhouses, with heating, basements and terraces if they should so desire. The only restriction is that the dwelling space of the main building may not exceed 50 m^2. It is even explicitly stated that all previous restrictions on erecting buildings are removed. Significantly, however, the notorious problems of obtaining building materials, seeds, fertilizer and other vital inputs are still referred to in traditionally vague terms of 'taking measures' in order to provide 'help'.[72]

As we have mentioned above, in official terminology considerable stress is placed on the 'collective' nature of activities in the gardening associations, but in practice it often tends to boil down to predominantly individual household gardening. The provisions made for *dacha* construction are only one illustration of the discrepancy between theory and reality. This focus on the individual, however, brings out a number of highly complicated issues regarding moral and legal rights to the plots of land that are involved. One of these has been dealt with in an article in *Literaturnaya Gazeta*, from which we shall draw some relevant evidence.[73]

The background story is as simple as the outcome is upsetting. After 36 years of service to his enterprise, a man dies and his son applies to become a member of the factory gardening association, in order to be able to 'inherit' the father's plot. This has been in the family's possession for a long time. The children have grown up there, trees have been planted and a small house constructed. Neighbours and other members of the association vouch for the family's good standing and record of hard work. No legal problems prevent such inheritance. The application, however, is denied and the plot is taken away from the family.

According to the 1966 model charter, this would probably not have happened, as such a decision would have been taken by the association membership itself. According to the revised 1978 version, however, the decision is to be taken jointly by the enterprise and the labour union, and the paper's correspondent, Kapitolina Kozhevnikova, argues that the new rules are 'directed at making the factory committees masters of the land, for the plots to be used as a tool in strengthening the cadres.' The excuse that was presented for removing the plot could only have served to add insult to injury: the adult son could not be considered a member of his father's family, since they had been living at different addresses. It is with considerable feeling that Kozhevnikova sums up the main issue involved:

> It is an age-old desire of man to organize his life, to provide a home for the family, it is a natural feeling. It exists in us as it does in nature itself. It is a complicated phenomenon, and in it are intertwined a multitude of problems – social and moral.

The underlying implications of this statement are rather provocative. Officially, as we have seen, the gardening associations are presented as collective organizations, which ought to serve to promote collective and socialist ideals and values, in accordance with the idea of a 'Soviet Man'. Here, however, are demands of a clearly individualist nature, which can serve only to counteract the official ideology, and Kozhevnikova makes no secret of her position: 'Morally as well as legally, the plot ought to belong to his children and grandchildren.'

Another reflection of the individualist nature of the gardening associations can be found in the problem of 'material supply', i.e. the provision of construction materials, tools and implements, fertilizer, seed, feed, etc. According to the model charter referred to above, such provision should be arranged ' "mainly" via the personal means of the members'.[74] There are of course 'recommendations' for enterprises and institutions to render assistance of various kinds, but actual experience informs us of the value of such promises of support. The state bank, for example, has instructions to grant loans of up to a thousand rubles, for five years, for the purposes of either buying or building a house, or for other improvements on the plot.[75] Similarly, enterprises and institutions have a *right* to use up to 25 per cent of their socio-cultural funds for investments in housing, roads, electricity, water supply, and other similar needs of the associations.[76] Fine as this may sound, however, it is hardly surprising that there are frequent complaints from those who are supposedly to receive the support in question.

In a survey of the experiences of the gardening associations, *Izvestia* presents a fairly bleak picture of the actual situation. First of all, it is obvious that the need for help is great. The alloted land is often remote, of poor quality, lacking roads, electricity and a reliable supply of fresh water. In addition, there are of course numerous problems with the quality and availability of tools and implements. (The latter will be discussed at greater length in Chapter 5.)

Furthermore, the attitude that is taken by officials in charge of providing help and assistance is predictably not very supportive. In one case, cited by *Izvestia*, of some association members turning to their 'parent' enterprise for the promised help, the message was spelled out rather clearly: 'For you we have no materials whatsoever. Forthwith you may as well save yourselves the trouble asking.' The conclusion that is reached by the paper is equally clear:

> The actual situation depends entirely on the initiative of the individual member. In the most incredible ways (partly, unfortunately, with the help of fixers and speculators) he is forced to arrange building materials, fertilizer, tools and implements, as well as seeds and plants.[77]

Shmelev also mentions that demand for garden plots is far from satisfied, with maybe only one out of every three or four families who want a plot actually receiving one. The usual excuse is of course a lack of suitable land, but he also finds that 'artificial obstacles of an administrative and bureaucratic nature are frequently constructed.'[78] Land, for example, is often alloted in remote locations, far from home and maybe even from major roads, this causing unnecessary loss of time. All of this should be seen against the background of the moral and educational role that is officially assigned to these allegedly 'collective' associations.

What has been said above regarding the gardening associations represents, nevertheless, the least complicated side of the private fringe. Let us now proceed to the more troublesome area – the *dachniki*. In so doing – it may be observed – we shall neglect the middle ground of the spectrum of activities performed on the private fringe, i.e. that of urban houseowners with gardens.[79] This neglect is motivated by the absence of such organizational and ideological difficulties that can be found at the respective ends of the spectrum.

The Troublesome Dachniki

Dachnik is a term that is frequently used with reference to urban residents who in various, unorganized, ways are engaged in agriculture. Literally, it refers to the Russian word for a small summer house – a *dacha* – and this gives a first impression of what we are dealing with here: people who visit relatives on the farms or in the villages, people who may have inherited a *dacha* and a small plot of land, or people who wish to purchase abandoned houses from *kolkhozy* or *sovkhozy*. In all three cases, the motives for wanting a plot are partly social and partly economic. They combine a desire for recreation, on the one hand, with the provision of an extra source of scarce foodstuffs on the other.

At first sight, it would be hard to see what it is that causes tempers to run high over the presence of such people in rural areas. Yet, from reading accounts in Soviet journals and newspapers, one is given a distinct impression of a highly controversial issue. This is particularly the case regarding such letters from readers that are regularly published in daily and weekly papers. We may also mention the explicit prohibition against admitting *dachniki* as members of the gardening associations that can be found in the model charter of these associations.[80]

The arguments that are presented in this 'debate' can sometimes be of a rather peculiar nature, and in order to provide a coherent point of departure we shall start by listening to the words of Aleksandr Nikitin, a correspondent of *Literaturnaya Gazeta* and a valiant champion of the private sector in agriculture.[81] Nikitin starts by referring to a letter received by the paper from a *sovkhoz* director in an area with a great influx of urban visitors.[82] The esteemed director is highly displeased with these *dachniki*, and presents a picture of their invasion which would probably not upset too many people in the West:

> At peak harvest time, they walk around the village in swimsuits, they lure tractor drivers into drinking, by offering vodka in return for various forms of help. And the working people cannot get to the blueberries in the forest, since they have all been taken by the people from the cities.[83]

The general impression of his complaints is that the urbanites are a disturbance, and he sees it as 'no coincidence' (*ne sluchaino*) that his *sovkhoz* had been showing poor results. Against this Nikitin retorts rather forcefully, starting with the observation that more than half the population in the village in question were pensioners, that many of the workers were close to pensionable age and that youngsters were few and far between. Consequently, many of the *dachniki* must be children and grandchildren of those who remain in the villages: 'What shall we do with them? Organize roadblocks to stop them from visiting their relatives?'

The moral side of the issue has its obvious points, but Nikitin also underscores its economic dimension, which is of considerable relevance for our discussion:

> The personal sector rests chiefly on the shoulders of old people, living in small and medium sized villages. Now tell me, would it be possible for these old folks to make such an enormous contribution to the national economy without the help of what they call '*dachniki*' – children and grandchildren from the cities? As we know, work on the personal plots remains largely manual.

Nikitin notes the rather obvious fact that townspeople go to the country not only for recreation, but also to work in fields and gardens. Their visits thus bring not only a valuable contribution of extra labour, which can be desperately needed, but there is also the added benefit of carloads of produce being brought back to the cities, thus relieving some of the strain on the official supply system. Why, however, 'do some officials focus only on blueberries, swimsuits and other such absolute rubbish, refusing completely to face up to this "social giant"?' The answer of course is that the private sector so far has not been included in the sphere of responsibility of such people, and if it should vanish altogether, this would simply mean, from their point of view, a greater supply of feed for livestock in the socialized sector.

Fortunately, such attitudes are not shared by all officials in responsible positions. Nikitin even refers to cases where contracts have been signed, promising the *dachniki* small plots of land, an abandoned house and help in various respects, in return for a share of their output and whatever labour they can contribute at times of peak demand. Many of these *dachniki* are of course pensioners, and many of them live for several months in the village. Some even make it their official residence (i.e. having arranged a *propiska*). Others are workers, who combine physical exertion and a healthy stay in the fresh air with bringing back scarce produce to the city.

An important fact to note is that none of this is actually in contravention of Soviet law. Houses may be bought – and inherited. Only land is excluded from such transactions. All land, including that *under* the buildings, formally belongs to the state, being placed at the long-term disposal of the farms, which may in turn come to agreements with prospective *dachniki*, should they have vacant land for such purposes.[84]

In many areas of the country, particularly in the non-black earth zone, villages are being depopulated. There are today about 725,000 abandoned houses in the Soviet Union as a whole (500,000 of which in the RSFSR), and about 200,000 hectares of abandoned but potentially fertile land.[85] Such villages have been termed 'futureless' (*neperspektivnye*) and a policy of resettlement, into larger and more viable rural centres, has been pursued, with much resistance and little success. Nikitin suggests a reversal of this policy. By forming 'Associations of Work and Rest' (*tovarishchestva trud i otdykh*), abbreviated as TTO, such villages could attract new hands to help out, and the urban population would be given a chance to enjoy healthy and meaningful vacations.

In summing up the proposal, the editors of *Literaturnaya Gazeta* asked for the views of their readers. Some months later another article was published, surveying some of the letters that had been received.

One important fact reported here was that about 60 per cent of the letters received were 'warmly' in favour of the idea of TTO.[86] There were, however, negative responses as well, and we shall continue by taking a somewhat closer look at both sides of this miniature 'debate'.

The first question is whether the *dachnik* is useful (*polezen*). Here the arguments are somewhat familiar. A certain V. Ivashchenko stands up in defence of the *sovkhoz* director who had been previously criticized by Nikitin, arguing that the attitude taken by the director was fundamentally correct, and that the presence of these *dachniki* undoubtedly does distract the *kolkhoz* and *sovkhoz* workers from their chores in the socialized sector. If empty land is subdivided into numerous tiny plots we are soon going to find the large fields empty, as the peasants are busy helping out on the plots, be it in return for vodka or whatever. As a matter of fact, the private sector as a whole – it is argued – forms a chain around the necks of *kolkhoz* chairmen and *sovkhoz* directors, who are faced with large empty fields and idle tractors as 'their' peasants are away growing strawberries on their tiny plots.

In contrast to this outburst there is a letter from another *sovkhoz* director, who is very pleased with 'his' *dachniki*. Having on his hands more than a hundred hectares of largely inaccessible, and thus useless, strips of land, he draws the seemingly obvious conclusion that if people from the cities are prepared to work these lands, to clear shrubberies and drain marshes, then this would present a clear gain for all. A minimum of some five hundred tons of potatoes could be grown, which would feed some five thousand people in town, and thus help depress prices on the market. Nikitin's suggestion on forming TTOs in the 'futureless' villages would be 'good for the members, good for the *sovkhozy*, good for the market, and good for society.'

What is hiding behind this exchange is to some extent a question of opportunity costs. It is certainly true that much produce is obtained from the various activities in the private sector, produce on which the country is becoming more and more dependent. At the same time, however, there is also a cost involved, a cost which may not be fully appreciated. The latter is manifested in terms of a fragmentation of the land as well as a diversion of resources from the socialized sector, above all labour but also machinery and other inputs. All of this contributes to the rapidly growing need for 'help' to be rendered by the urban sector, a form of assistance which has its obvious price in terms of industrial output foregone. Some of the more heated contributions to the debate on this issue reflect the fact that the relative magnitudes of costs and benefits will vary greatly between different parts of the country, depending on the local conditions for plot agriculture. It is one thing, for example, to grow citrus fruit in Georgia, and quite another to grow cabbage in Tambov.

The second question raised concerns the motivations behind the urban population's longing for the land, and here the temperature of the debate increases considerably. Under the heading of 'these devious urbanites', a certain B. Roskov stresses the need to make a distinction between granting small plots of land to pensioners in rural areas, and the growing practice of allowing urban residents to buy houses in the villages, and thus also to lay claim to plots of land. According to the principle of 'give him an inch and he will take a mile', a picture is presented of how these people, who are contemptuously referred to as *chastniki* ('privateers'), will gradually take over: 'It starts with a custom built house, then there are demands for enlargement of the plot, for fancy wells to be drilled, for electricity, for their land to be ploughed up, and all to be done by the *kolkhoz*.' The argument that help is thus provided towards solving the food problem is arrogantly brushed away. It may well be that some help is provided by these urban pensioners, but in the future, when production in *kolkhozy* and *sovkhozy* runs smoothly, the *dachniki* will give rise to a new and more serious social problem. According to Roskov, the problem of vacant lands in the 'futureless' villages must be solved in cooperation with state enterprises and institutions, for them to form *podkhozy* and maybe even pensions for vacationers.

A completely different aspect of the longing for the land is presented by a certain M. Antonov, in a letter with the heading 'No Whim – a Demand of Our Times'. To Antonov, the questions raised by Nikitin reflect 'one of the most pressing problems of contemporary civilization.' The time has long passed when the cities were looked upon as paradise. People are tired of crowds, asphalt, odour and noise, and the chance to spend a month in quiet surroundings, close to nature, would surely be welcome. The TTO is one way of satisfying such needs. Moreover, it would not only offer a healthy recreation for people from the towns, but would also contribute to erasing the differences between town and country. By integrating the urban population into agriculture a new organizational form would arise, a 'workshop of health' (*tsekh zdorovya*) which would be clearly superior to the present practice of 'drafting' urban people to help out at the busy periods. The TTO would be a 'progressive' way of 'liquidating' not only differences between town and country, but also 'the unnatural crowding of people in gigantic cities.'

The controversy that is involved on this count is perhaps even more clear cut than in the previous case. The urban population would surely appreciate being offered plots and houses in rural areas, for reasons outlined above, and at first glance it would seem hard to defend the hostile attitude that is taken by Roskov. If, however, we look more closely at his arguments – against the very special background provided by Soviet agriculture – they begin to raise more serious implications. The problem of fragmenting resources is fairly straightforward, and has been dealt with above. Of greater importance, however, is the problem of changes in the social structure of the villages. As Yanov has shown, it was largely this type of problems that put an end to those experiments with 'normless teams' (*beznaryadnye zvenya*) in Soviet agriculture which have been mentioned above. Although superficial at first glance, this argument may thus have a serious content, and it will be returned to later.

The third question raised concerns the need for organization, and here we are treading on highly controversial ground. N. Romanov, a former *kolkhoz* bookkeeper, argues that all those people from the cities who already have small plots in the villages would be strongly against this idea of a TTO. Why should the *dachnik* submit to such a new form of 'serfdom' (*barshchina*) that is implied by signing contracts to hand over part of his harvest, and to perform labour on the fields of the socialized sector? The only result would be that he would flee into the depth of the woods, 'far from the jealous organizers of his country rest.' Or he would join the *sovkhoz*, roll down his sleeves, mark time and have a generally demoralizing influence. The collective is a good thing, but at times even here, one needs to take a rest: 'What we need is not TTOs, but a mass distribution of plots of land to all interested citizens. The plot should be a legal right for anybody with a house, be they of urban or rural extraction.'

Against this anarchistic vision, it is rather fitting that the paper allows an officer of the reserves to conclude that 'without the collective we are privateers (*chastniki*)'. The soil is a gift, which should not be handed out left, right and centre. But it is a gift only when cultivated, not when left to weeds and scrub. Everything grown is a contribution towards solving the food problem, so why not allow the urban population to take part? The formation of TTOs, or of some such organizations, would preserve the collective nature of work and preclude the emergence of 'privateers'.

The issue at stake here lies at the very heart of the ideological justification for 'personal' activities. These can be accepted as long as it can be convincingly argued that they are not independent ventures, but rather form an integral part of the overall socialized sector. The demand made by Romanov, for plots to be distributed without any organizational 'cover' whatsoever, is thus a considerable challenge to the entire legitimacy of 'personal' activities as such, and the fact that it was published gives an indication of the ambivalence of the present official attitude.

In a third article, Nikitin returns to survey some further correspondence received, and partly perhaps also to defend his original proposition on the TTOs.[87] The latter presumption is derived from the following, rather defensive outburst at the beginning of the article:

> It is not necessary, comrades, to knock down open doors, to prove to your opponents that the pensioner-*dachnik* does not feed the entire nation, that the villages need young people, developed rural centres, and roads. The question is simpler than that: is work performed by the urban people a help or an obstacle?

To the latter question Nikitin claims that the majority of letters received had indicated that help was provided. About 39 per cent were in favour of the TTO, as proposed. Another 40 per cent were in favour of urban contributions, but under different forms, while only 14 per cent were against all forms of *dachniki* and *chastniki*. Numerous excerpts are quoted, from letters received, and suitable responses given where the opinions expressed are deemed to be 'incorrect'. For example, the frequently used argument, that it is a lack of suitable land that slows down the development of the private fringe, is refuted in a rather forceful manner:

> No, it is not a lack of land that is acting as a brake on the promotion of the movement of 'city-peasants'. Nor is it legal or economic problems. It is simply a distrust of this movement, it is a fear of various social distortions – that more than anything else is what is holding it up.

The struggle that has been waged by Nikitin and others is no doubt an important one, and there are certainly great numbers of urban residents who root for them. It is also indicative of Gorbachev's *perestroika* that a decree was issued recently, permitting urban residents to buy abandoned houses and to cultivate adjacent plots of land.[88] At the same time, however, we must recognize that resistance from members of the controlling bureaucracy is not only widespread but also – as Nikitin points out – derived from fears of developments that, from their point of view, might be seriously disruptive. Consequently, as late as August 1987, i.e. in the midst of Gorbachev's *perestroika*, the following could be read in *Sovetskaya Rossiya*:

> In Moscow alone there are a million unsatisfied requests for garden plots [*sadovye uchastki*]. Less than 10 per cent of the urban households, in the capital as well as in the provinces, achieve this dream. People from the city rush into the countryside for the growing season. They are ready to take on homes. They are ready to cultivate the adjacent land. To feed their families. To offer their

> surplus produce for sale. But can we give them an answer? No, it is not useful, forbidden, we will not allow them to take the land.[89]

This fairly pointed statement concludes the presentation of the *dachniki*. The findings thus far can be summarized as follows.

Conclusion

This chapter has presented a picture of the two supportive fringes of Soviet agriculture. In accordance with the previous presentation of Soviet agriculture as a Black Hole, we have seen here how rapidly increasing numbers of both industrial enterprises and members of the urban population have been mobilized to compensate for the losses in output that result from poor management and sloppy work in the socialized sector of agriculture.

Although such help may certainly at first glance seem to be needed, we have tried to cast some doubt on the efficiency that characterizes these operations. We have seen that the performance of the *podkhoz* sector leaves quite a lot to be desired, and that the gardening associations and the *dachniki* as well are engaged in fighting an uphill battle. The perhaps most serious indication of something being amiss here is formed by those processes of 'reversal' that we have observed on both fringes, i.e. how industry, on the one hand, is becoming increasingly involved in agriculture, whereas the farms are setting up makeshift workshops of their own. On the one hand, the urban population is increasingly turning to the soil, while on the other hand, the rural population is turning to the cities.

We shall expand on all of these points at greater length below. Chapter 4 will continue by presenting those views for and against private activities that are voiced in the Soviet debate. Chapter 5 takes a closer look at the problem of material supply for the private sector. The really crucial aspects, however, will be dealt with in Chapter 6, where we shall argue that the costs of the support that is rendered from the fringes may in the long run turn out to be of an unexpected nature as well as magnitude. The latter will be the topic for discussion in the concluding chapter.

Chapter four

Soviet attitudes to the private sector

Attitudes to the private sector are marked by much the same conflict that underlies the puzzlement expressed at the outset of this study. On the one hand we find a clear realization that the private contribution to overall food supply is of such a magnitude that the country really cannot do without it, at least not in the near future. This realization is frequently encountered in Soviet sources, and will also figure prominently below. For the moment, the following statement by economists G. Dyachkov and A. Sorokin can be considered to be representative:

> The process of 'dying out' of auxiliary agriculture amongst all groups of the population cannot be the consequence of administrative measures, but only of an increased output from the socialized sector in agriculture, of increased incomes from that sector, of the creation of a reliable rural trading network, and of a reconstruction of the existing system of rural housing.[1]

What is involved here is precisely that hope which was at the heart of Khrushchev's policy towards the plots, i.e. that the socialized sector should have become strong enough to shoulder the responsibilities of the private sector, and that the rural infrastructure would have reached a level of development sufficient to arrest the flight of the young to the cities. In view of Soviet agricultural performance since Khrushchev's time, however, the attainment of this hope seems to be a tall order indeed, indicating that we may be destined to live with the plots for quite some time to come.

On the other hand, there is the view that this ought not to be so, that this private island inside the socialist economy has a highly unfortunate influence on the builders of socialism, in moral as well as in ideological terms. This view can be illustrated by a letter sent to *Literaturnaya Gazeta*, in contribution to the debate on the role of the urban participants – the *dachniki* – that was discussed in the previous chapter:

> Our national agriculture should not depend on the help of *dachniki* and other private contributions. It needs young, educated people, freed from the necessity of tilling their own plots, people who can devote their time to books and television.[2]

Not knowing the special peculiarities of the private sector, it would be hard indeed not to sympathize with the author of these lines. Moreover, it again raises a question of crucial importance regarding the *actual* role that is played by the private sector. Is it a valuable support for the socialized sector, or is it an obstacle on the road to 'perfection' (*sovershchenstvovanie*) of socialized agriculture? As we have indicated above, this question was central to Khrushchev, it was at the heart of Raig's vision of the future, and it was even raised by Nikitin, in his final summary statement on the TTOs and the potential usefulness of the *dachniki*. Needless to say, it will also have an important bearing on the discussion in the remaining chapters of this study.

The present chapter takes a closer look at various arguments presented in the Soviet debate regarding the general usefulness of the private sector. The questions of support or obstacle will figure prominently here. We shall look at arguments that concern both the purely ideological aspects of private activities in the midst of building socialism, and the perhaps even greater problem of popular discontent with alleged 'speculation'. Moreover, we shall extend the scope of the discussion to cover not only the various plots, but also the *kolkhoz* markets. The latter of course form the real breeding ground for speculation – perceived as well as real – and will thus have to be explicitly considered.

The picture that is thus assembled will then be used as background for our subsequent discussion of the formation of policy towards the private sector in general, and of the behaviour of local officials in particular. The discussion starts by looking at the ideologically important issue of 'private' versus 'personal'.

Is It Really Private?

The seemingly amazing differences between input and output on the private plots are probably amongst those Soviet statistics that are most commonly cited in the West. If it really were the case that the socialized and collectivized peasants, in their free time, and with few other tools than the hand hoe, had managed to produce more than a quarter of the country's total agricultural output, on merely 1.5 per cent of its agricultural area, then this really ought to say something important about private versus socialist agriculture. Just imagine, for example, a fourfold increase in the total plot area. Wouldn't this then cover the country's total food requirements, making possible a total abolition of all state and collective farms?

For those Soviet economists who stand up in defence of the private sector this line of argumentation is particularly irksome, since it apparently plays an important role for 'domestic consumption' as well. Let us start by listening to Shmelev, the main champion of the plots, who pointedly formulates a question that perhaps ought to have led off this study:

> Do we support the LPKh only because it is at present economically indispensable (i.e. for simple pragmatic reasons)? Or is it a matter of principle as well, because this type of activity forms an integral part of socialist agriculture and the relations of production in the LPKh enter into the overall system of socialist relations of production?[3]

Which indeed, is the answer to that question? It is not hard to guess what Shmelev would like to suggest, but before entering into that discussion it may be of some interest to note that he considers the issue to be of particular importance because the LPKh 'forms a favourite subject for the inventions of bourgeois propagandists, seeking at any cost to defame socialist transformation of agriculture.'[4] A particular villain in this drama is the 'famous sovietologist' A. Nove of Glasgow, whose talk about a 'private' sector in Soviet agriculture presents a 'twisted' picture of socialist transformation and of the Party's agricultural policy.[5]

Shmelev's way of rebuffing the work of Nove and other 'apostles of anti-communism' forms another interesting point that we shall have reason to return to below. He argues that *if* the private sector could really be considered 'private', then this would not only testify to the unusual staying power of individualist traditions, over six-odd decades of Soviet power, and more than half a century after collectivization. It would also call into serious doubt the very idea of a socialist transformation of agriculture, since we would then be dealing here with 'millions of private entrepreneurs, having treasured deep in their souls individualist interests and leanings, in spite of the work of *kolkhozy* and *sovkhozy*.'[6] Again, Shmelev's conclusion ought to be obvious and again we shall have reason later to take issue with him. Suppose that it is the *burzhuaznye apologety* who are right?

Irrespective of the answer to that specific question there are, as we pointed out at the outset of the study, several good reasons for making a distinction between 'private' and 'personal', reasons which are chiefly linked with that strong interdependence between the plots and the socialized sector which we have touched upon above. We shall now proceed to take a closer look at these reasons, while recognizing that there are also several bad arguments used – for the same purpose – which must be refuted.

In his treatment of the issue of private versus personal, Wädekin has produced out of Soviet writings on the topic a list of five 'typical' lines of argument. These are reproduced below, in a somewhat abbreviated form:

1. One class cannot at the same time belong to both a socialist and a private sphere of production;
2. According to established rules, a plot cannot be operated unless its owner also participates in public work;
3. The plot becomes a problem only when it provides its owner with more than a subsistence minimum;
4. The peasants depend partly on the plot for their survival, or at least have done so until recently;
5. Operation of the plot is impossible without material support from the socialized sector.[7]

The first of these points is rightly dismissed as 'declamatory', but we shall disagree somewhat with the dismissal of both the second and the third as 'unconvincing'. Wädekin's focus is predominantly on the economic aspects of the problem at hand, and in this perspective it is certainly correct to say that simply setting normative limits adds nothing to the qualitative aspects of the matter. Here it is points four and five that are relevant, as they focus on the *nature* of the productive process. From our perspective, however, which focuses more on socio-political issues, such as attitudes and behaviour, the normative aspects involved are of some considerable importance. We shall return to them in a moment. In particular, we shall underline the importance of the argument on 'too much', i.e. the thin line beyond which legitimate personal support activities are transformed into activities which are branded as private 'speculation', without any actual change taking place in the *qualitative* nature of the productive process.

Before proceeding to that discussion, however, there is a rather important terminological point which needs to be dealt with. Soviet writers who are sympathetic to the private sector often emphasize that there is a marked interdependence between the private and the socialized sectors in the sphere of production. The nature of this interdependence can be summarized in the following three points. First, with the exception of pensioners, children, invalids and mothers with many and/or small children, all of those who are engaged in plot agriculture do so in their spare time, in addition to working full time for the socialized sector (in theory at least). Second, the plots are crucially dependent on the socialized sector for feed, seed and other mechanical or chemical inputs, and third, the socialized sector is in its turn dependent on the plots for various labour-intensive operations, such as the fattening of animals.

Seen from a strictly economic point of view, it would thus make sense to say that we are dealing here with an integrated whole, rather than with two competing types of agriculture, one being 'private' and the other 'socialized'. The highly different nature of production in the respective sectors makes it natural that a certain amount of interaction will arise between the two, and the plots should consequently be seen neither as an independent and alternative form of economic activity, nor as a mere appendix to the socialized sector. A more correct view would be that they serve the dual purpose of being at the 'personal' disposal of the households, for the bulk of their food needs, as well as performing an important support for the socialized sector in production.

It is, however, a matter of more than semantic importance whether we choose to talk about the economic interaction that takes place between the two sectors in terms of 'dependency'. From a purely ideological point of view, it is rather obvious that Soviet writers sympathetic to the private sector will wish to do so, simply in order to refute allegations of 'petty-bourgeois' tendencies. We shall return to this at great length in a moment. From a more narrow economic point of view, it is on the other hand equally obvious that such arguments do not make much sense. There is, however, a further dimension of the issue which does make it warranted to speak of dependence.

Since livestock production is an important part of private output, and since the households are not able either to produce or purchase from the outside sufficient quantities of feed, they are in a very real sense 'dependent' on the good will of farm managers and of local Party officials in this respect, just as they are 'dependent' on such authorities for the allocation of seeds and fertilizer, and of tools and implements. In a similar vein, the socialized sector is of course 'dependent' on the readiness of the peasants to perform good work for the collective. The main point of this argument is that we are dealing here with an economic relation – between the peasants and their 'prefects' – that is of potentially great mutual benefit, but which traditionally has been characterized on both sides by a very deep sense of distrust, if not outright hostility. This point is of central importance to the further presentation.

As far as the relations of production are concerned, we would thus be inclined to agree with making a distinction between the two sectors. This, however, is not the real core of the matter at hand. It is in the domain of the moral and psychological impact on the individuals concerned that the labels of 'personal' and 'private' become of paramount importance. Here is Shmelev again:

> The LPKh has not strengthened the private property instincts of the peasant. On the contrary, by means of that organic coordination of public and personal interests which has facilitated a socialist transformation of agriculture, it has been possible to overcome such attitudes. The destruction of private property instincts amongst the *kolkhoz* peasants, and their transformation into worker- collectivists, has been achieved not by changing their relations to the LPKh, but by changing their relations to the socialized sector.[8]

What is argued here by Shmelev is, in essence, not only that there is no contradiction between the two modes of production, but even that the strong interdependence between the two sectors has contributed towards the formation of a collectivist mentality among the peasantry, a mentality which is conducive to the ideals and values that underlie the official presentation of Soviet agriculture – as it *ought* to be. From our presentation above it should be obvious that we do not agree with this view, and the remaining parts of the study will be devoted to explaining why. An issue of pivotal importance in this context is the notion of 'private property instincts' (*chastnosobstvennicheskaya psikhologiya*). It figures prominently in the Soviet debate, and given the present focus on individual behaviour and attitudes it is obviously of great relevance for this presentation as well. We shall thus have reason to return to it at repeated occasions below. Let us proceed now to look at some of the components of the case that Soviet sources bring against the private sector.

In general terms, this can be said to be strongly focused along two lines: on the one hand such morally and ideologically harmful influences on individuals that allegedly follow from this type of 'petty bourgeois' (*melkoburzhuaznye*) activities, and on the other such 'speculative' gains that are derived from these activities. Our discussion below shall reflect this bifurcation, for the simple reason that the two types of attitudes will be found among very different groups of people, and will thus have different consequences and implications. While complaints about speculation are typical of people who are subjected to high prices on the *kolkhoz* markets, or who see the rural *nouveau-riches* driving around in new cars, the more theoretical type of arguments will be voiced largely by economists and policymakers. Let us start by looking at the latter.

Moral and Ideological Aspects of 'Private' Agriculture

At a round table discussion on the theoretical and practical aspects of supporting the private sector, which was held in Moscow in April 1981, academician A. A. Nikonov of the VASKhNIL, the Lenin Academy of Agricultural Sciences, summarized the role of that sector under the following three headings. First, it has an *economic* function, in supplying both the rural and the urban population with a range of foodstuffs. Second, it has a *social* function, in providing the rural population with an additional source of income and the urban population with the possibility of meaningful recreation. Third, it has an *educational* function, in socializing children and youngsters to sound working habits.[9]

Each of these three points presents a rather complicated problem of its own, which certainly merits separate treatment. For the moment, however, we shall focus exclusively on the economic dimension. The educational function will be dealt with in greater detail below, in relation to the discussion of peasant response to the various forms of discrimination and harassment that follow from the attitudes described in the present chapter. It will thus only be referred to very briefly here. The social function has, on the other hand, already been referred to above, in relation to the mobilization of additional labour, and the recreational aspects of the TTOs and the *dachniki*. It will consequently be discussed only *en passant* in the following chapters.

In order to economize on space, we shall start the investigation of the ingredient parts of the economic case by making use of the presentation in the article by Dyachkov and Sorokin that was quoted at the outset of this chapter. Here the authors, who are obviously sympathetic to the private sector, have compiled a veritable catalogue of opinions expressed by people against whom they apparently wish to polemicize.[10] We shall use this as representative of the main arguments that are usually presented by Soviet writers on the topic.

For the protagonists of the private sector in agriculture, the one really crucial undertaking is to repel ideological flak by establishing that the plots in no way represent a *qualitatively* different, i.e. essentially capitalist, mode of production. Instead it has to be argued that they form an integral part of the overall – socialized – agricultural sector, a part which in various ways acts as a support for rather than an obstacle to the development of that sector, and which thus serves to promote rather than obstruct the building of socialism. The consistently made distinction between 'private' and 'personal' is an integral part of this undertaking, and it is a tell-tale sign of hostility towards the private sector when the term *chastnyi* (private) is used, in lieu of the more 'correct' term *lichnyi* (personal). Given this ambition, it should not be surprising to find references to the opposition where one-line statements are quoted, preferably out of context, in a way so as to present a strawman-type target for arguments intended to establish the allegedly true nature of the plots. Such, anyway, is clearly the strategy in the article at hand, and we shall thus have to exercise some care in disentangling the respective arguments that are put forward.

An important bone of contention is of course the justification of small-scale private plots in the environment of a large-scale socialized and mechanized agriculture. At this point, two of the 'opponents' chosen by Dyachkov and Sorokin – sociologists M. N. Rutkevich and F. R. Filippov – are quoted as being of the opinion that the private sector 'forms a remnant of private, largely subsistence oriented production, the main aim of which is to cater to the household's needs for a number of foodstuffs.'[11]

The quote is taken from a 1970 book on social transformations, and the context is that of social change in the villages. The source can probably be considered as quite representative of the orthodox sociological stand of viewing all forms of private undertakings as suspect. It might, however, have been reasonable also to include the following fairly non-ideological statement:

> Such undertakings [i.e. private plot activities] will influence peoples' lifestyle: cultivation of the soil, husbandry of cows, hogs and poultry, and the need to plan for winter supplies, all of this will leave an imprint on the peasant's way of life as well as on his psychology.[12]

In the discussion by Dyachkov and Sorokin, however, no interest in matters of this kind can be perceived. Instead, the authors, who apparently feel that they have an axe of a different kind to grind, proceed to gather more evidence of hostile attitudes to the plots. To give but one example, they quote a standard 1963 Soviet textbook on political economy, which presents the private sector as 'a brake on the road to a complete removal of the remnants of private property instincts amongst the peasants-*kolkhozniki*.'[13]

This limited scope is unfortunate, since a broader perspective obviously would have been of great relevance for a discussion of the educational function of the private sector, as indicated by Academician Nikonov above. Fortunately, it is not only Rutkevich and Filippov who recognize that private agricultural activities, pursued under self-determination and for private gain, may have a detrimental influence on the mentality of the alleged builders of socialism.

In a study of the influence of the 'Scientific-technological revolution' on class structure, for example, L. S. Blyakhman and O. I.

Shkaratan point out the fact that the working class preserves within itself certain 'remnant relations of a pre-socialist nature', one example of which is the private plot. With due reference to Rutkevich, the authors interestingly emphasize that the possession of plots, which forms a main socio-economic difference between town and country, may 'have a serious influence on certain aspects of the lives of workers who are engaged in such activities.'[14]

In their view, the plot serves as a potential source of moral and ideological contamination, a stand which contrasts rather forcefully with the ideologically supportive functions ascribed to it by Shmelev above. These arguments are obviously relevant to the discussion of both the private sector at large and perhaps even more so to the private fringe and we shall thus return to them at greater length below. For the moment, however, let us return to the line of argument pursued by Dyachkov and Sorokin will be followed.

Although the sources cited may perhaps not be representative in terms of reflecting recent developments – in either reality or debate – they do reflect the apparent permanency of that fundamental dilemma of the private sector which was indicated at the outset of this chapter. If the Soviet Union is to have an efficient modern agricultural sector, worthy of an industrialized superpower, one would certainly be inclined to agree that the plots act as a brake on development, in terms of both a fragmentation of resource use and the undermining of incentives to work for the collective. Whether we choose to discuss this in terms of petty-bourgeois leanings and remnants of capitalism is a matter of taste. The fundamental conflict, however, should be as obvious today as it once ought to have been to Khrushchev.

This central dilemma regarding the future of the private sector is also dealt with by Rutkevich and Filippov. On the one hand, they acknowledge that on two occasions during the post-war period agriculture has been caused considerable harm by 'voluntaristic errors in theory and practice' which produced 'attempts at artificially speeding up [the private sector's] process of dying out', while on the other, they unequivocally establish the same long-term goal as that presented above:

> As plot agriculture becomes economically unattractive to the rural population its extent will inevitably be reduced, initially in relative terms but over time also absolutely. This reduction, however, must be the result not of administrative measures but of economic necessity.[15]

In this perspective, the defence of the plots put forward by Dyachkov and Sorokin has a rather peculiar ring. Instead of discussing the nature of, and possible prospects for, production in this sector as such, the authors simply concentrate on demonstrating that it cannot be considered as an independent private (*chastnyi*) sector. While a 'private' peasant in the West incurs large expenditures for land, construction, tools and implements, etc., 'personal' agriculture in the Soviet Union takes place on land owned by the state and benefits heavily from support rendered by the socialized sector. Moreover, since work performed by a private peasant is not planned and organized, like that of his Soviet counterpart, the former alone can be referred to as private (*chastnyi*) work. We have here a quite explicit formulation of the ideological understanding of interdependence.

Carefully avoiding any mention of possible social or psychological influences on the individuals involved, a number of similar points are made, all of which allegedly lead to the conclusion that the risk of a capitalist transformation of the private sector, into market-oriented production, can be firmly ruled out. Consequently, the opinion, quoted from a 1956 book on *kolkhoz* production, that 'in every form of personal activity there lies imbedded the possibility of a capitalist transformation',[16] is branded as 'untenable'.

As we shall see in the latter part of this chapter, the risk of capitalist contamination is considered to be particularly great with respect to trade on the *kolkhoz* markets. In order to tackle this problem, Dyachkov and Sorokin quote another source which allegedly holds that such trade might lead to 'the growth of elements of spontaneity in the national economy, i.e. in fact a danger to socialism.'[17] 'Spontaneity', of course, is a classic Soviet pseudonym for market relations and it is far from unusual to hear it used with respect to the *kolkhoz* markets. The choice of source used to document this allegation, however, provides a real gem in the art of making misleading quotations.

Not only is it incorrect, since the latter half of the sentence has been freely added by the authors. It is also grossly out of context. The book from which it was lifted was published in 1953 and deals with property in socialist society. The original context is that the rapid development of *kolkhoz* production will lead to increased sales on the *kolkhoz* markets, at freely formed prices, a development which threatens to produce the above 'elements of spontaneity'. The solution to this dilemma, moreover, which Dyachkov and Sorokin have not found it opportune to quote, is provided by none other than Comrade Stalin himself. In his famous 1952 textbook on the economic problems of socialism in the USSR we can read the following:

> In order to raise *kolkhoz* property to the level of national property, it is necessary to exclude surplus *kolkhoz* production from the system of trade, and relegate it to direct product exchange between state enterprises and *kolkhozy*.[18]

The present author may perhaps be forgiven for questioning the relevance of Stalin's views on *kolkhoz* trade in the 1950s to issues regarding the current position of urban and rural families engaged in private plot activities. Yet, one might possibly suspect a message written between the lines here, i.e. an attempt to associate those antagonistic to the private sector of today with the protagonists of Stalinism. Be this as it may, it remains a fact that the subsequent line of argument is consistent with that 'model' strategy which was hinted at above, i.e. to erect ideological strawmen that are easily overthrown.

Thus it is claimed that 'in a number of works', one can encounter the idea that the growth of the private sector is based on remnants of 'petty bourgeois' and 'private property' attitudes amongst sections of the population. There is only one source advanced, however, in order to document this statement, namely that same book by Rutkevich and Filippov which was quoted above. We are beginning to discern who the main enemies are. Dyachkov and Sorokin then proceed to refute the allegations by using arguments similar to those presented in the above discussion of private versus personal, i.e. largely normative statements on the nature of private production.

Again, the terminology is of course somewhat beside the point. We do, however, find it hard to believe that an expansion of the scope for private activities, under self-determination and for private gain, should be void of any contradictory influence on the parallel attempts at creating a collectively minded Soviet Man, ready to work for the common good rather than for his own. It would seem harder still to believe that such an influence would actually serve to strengthen the collective consciousness of those involved, as was argued by Shmelev above. We shall return to these central issues at greater length below, but in order to show that the arguments raised above should not be seen as a special case, let us even now listen to a more recent exchange, between two of the main combatants involved – Rutkevich versus Shmelev.

Our understanding of the arguments that have been outlined above as a catalogue of typical views is supported by a 1985 article in *Kommunist*, the ideological organ of the Central Committee, where the very same Rutkevich takes a markedly hard-line stand, repeating *con brio* virtually all the arguments in our 'catalogue'.[19] His focus is on the development of Soviet society towards a 'classless structure'. In this perspective, the private sector in agriculture is characterized in a way that must certainly act as a provocation to people like Dyachkov and Sorokin, or indeed to Shmelev:

> Thus, under present conditions personal auxiliary agriculture presents itself quite specifically as a *remnant form of small scale*

> *production*, which may be *partly* characterized as private, depending on the extent to which its output acquires the form of a good.[20]

There are two words used here – 'private' (*chastnyi*) and 'good' (*tovar*) – which both indicate a hostile attitude to the private sector, in that it is attributed to be of a capitalist nature. Rutkevich continues his presentation in the same vein, by arguing that support for the private plots must be accompanied by a realization of their inherently negative tendencies: 'Individual activities will in many cases lead to the preservation, among part of the population, of remnants of private property instincts.' The latter argument should be familiar from the discussion above.

He then goes on to claim that particular attention should be given to such cases where work on the plots is turned into a main occupation and a main source of income. Benefiting from free or heavily subsidized inputs, such as land, electricity and feed, the plot is accused of being transformed into a small-scale private (*chastnoe*) enterprise, 'specializing' in early vegetables, flowers, etc., which are sold at 'speculative' prices on the market. Against such 'harmful tendencies' a struggle must be waged, which 'requires no little effort.' Finally, it may be noted that Rutkevich also considers it 'inevitable' to undertake 'certain changes in the rules governing market trade.' No doubt, the latter amounts to a call for regulation of prices that are at present not subject to legal controls.

Shmelev replied to this attack some weeks later in *Ekonomicheskaya Gazeta*, expressing surprise that such views can still be found 'in our press'.[21] Rutkevich is accused not only of ignoring the findings of economic science, but also of adopting a clearly unscientific method of analysis. The latter is rather amusing, considering Shmelev's own way of refuting Rutkevich's allegations of the capitalist aspects of plot production. First, he quotes parts of that same passage about 'remnant' characteristics, and of production acquiring the form of a 'good', which was presented above. The message intended by Rutkevich is of course that marketing, for private gain, of privately grown produce will have an influence on the mentality of the producer, but Shmelev chooses an entirely normative defence which totally ignores the really crucial issue of psychology.

The form of a 'good', it is argued, is acquired by output from small-scale peasant and handicraft enterprise, and from capitalist firms. The same, however, is also true for output from socialist state and cooperative enterprises, and according to the logic of Rutkevich's argument all these diverse forms of production thus ought to share the same 'characteristic private [*chastnyi*] appearance'. In conclusion,

Rutkevich himself is given a good brush-off: 'In relation to such a "theory", one can only repeat that it is quite, quite strange that it finds expression in the pages of the press.'

We shall refrain from passing judgement on the arguments presented in this debate, but it is of some importance to note that at present it is the proponents of the private sector who have the upper hand. This impression was brought home quite forcefully at a round table conference on problems in agriculture, held in Moscow towards the end of 1986, under the auspices of the journal *Selskaya Nov*. At this conference, L. Nikiforov, head of the Institute of Economics at the Academy of Sciences, summed up the current situation in noting that several recent Party decisions had called for a stand to be taken against a number of old stereotypes:

> One such stereotype concerns the attitude to personal auxiliary agriculture. Briefly put, this stereotype can be characterized in one single word – 'privateer' [*chastnik*]. This word, and all its associations, continues to have a negative influence on the development of the LPKh. It has gained currency partly with the help from certain scientists, the mass media, and from practical workers, particularly at the *raion* level, as well as occasionally from workers in agricultural enterprises.[22]

Somewhat further on in his speech, Nikiforov returns to that same theme by stating that the private sector 'does not constitute a separate individual sector or a manifestation of private property instincts, as it is sometimes – unfortunately – thought.'[23] Not only do these statements constitute sharp signals in the direction of those who have previously been critical of the private sector. It is also interesting to note that Nikiforov uses the Russian expression *individualnyi sektor* (individual sector), rather than either of the two previous opponents *chastnyi* (private) or *lichnyi* (personal).

This novelty may be interpreted as an attempt to create a fresh start, rather than to side openly with the previous protagonists of the plots, and it is interesting to see that same expression used in the September 1987 decree on agriculture.[24] It is important to note in this context, however, that while negative attitudes towards private agriculture are presently largely absent from the press, they can certainly not be assumed to have disappeared from the minds of those who have previously been harsh critics. A reversal of the official attitude may thus produce a very sudden and dramatic change of stage.

Before concluding the discussion of the moral and ideological aspects of the private sector, however, we shall also take a somewhat more detailed look at the issue of the *kolkhoz* markets, referred to above as a particularly sensitive ideological ground. Here is found an identical line-up of pros and cons as in the case of the plots. Some people simply seem to be convinced that such trade that takes place on the markets really *should* be handled by the official system. Accordingly they argue that the former acts as a drag on development. This classic argument has been around for a long time and it may well come to outlive the current wager on the plots. In his study of the private sector, for example, Wädekin offers the following illustration, from *Ekonomicheskaya Gazeta* of 1965:

> Some employees of state trade organizations and of the city soviets believe that [the *kolkhoz* market] is a relic of the past, that the markets compete with state trade, and that they should be pushed aside. On the basis of such false assumptions, they usually select sites for the markets in the older, outlying fringe areas of the city, so that townspeople are at times simply unable to get there.[25]

What we have here is a clear case of self-validating ideological beliefs. On the one hand, the markets cannot be completely suppressed, since that would have disastrous consequences for the country's food supply. On the other hand, they are not allowed fully to develop their potential efficiency, and will thus - by necessity – come to acquire a 'backward' appearance. The latter will then serve to strengthen the ideological case against their existence. The problems of poor support, that follow from such attitudes, will be discussed at greater length in the following chapter, but before proceeding to that discussion let us listen to a recent voice, which underlines the continuity of those attitudes that have been expressed above.

In 1979, i.e. following the 1977 decree on support for the private sector, Shmelev surveyed the situation on the markets, noting both the existing problems and the official determination to work in the direction of improvement, by rendering help and support of various kinds. His conclusion, however, is not very encouraging:

> It would all seem to be very clear. Nevertheless, there are people, some of them in high administrative positions, who think that under present conditions the market has outlived itself, and that there is thus no need to pay any attention to it.[26]

Two important – and serious – problems can be identified here. The first concerns the time that is expended by those hundreds of thousands of people that do the actual trading on the market floor, while the other focuses on the time that is lost on the road, taking the produce to market. As Dyachkov and Sorokin note, information on the magnitude of such losses is certainly difficult to obtain: 'Unfortunately, the statistical authorities do not compile data on labour expenditure and thus it is not possible to analyse its development over time.'[27] There does exist some scattered data, which provide a rather sombre impression. According to a source from 1968, some 600,000 people were engaged in trading on the market. This involved the expenditure of more than 200 million man-days.[28] In 1979, Shmelev wrote that we are dealing with a few hundred thousand people, expending annually a similar amount of time as in the 1968 case.[29] The latter figure is repeated in the following year by Voronin, in an article about the role of trade to the plots.[30] The problem with the latter sources, however, is not only that it is unclear whether they refer to time lost in both transport and trading, or to only one of the two. More serious is that they seem to rely on the same 1968 estimate. What the actual magnitude of the problem is today nobody seems to know.

All in all, it is fairly obvious that a considerable price is paid for the function that is performed by the markets. As is often pointed out in the Soviet literature, it is chiefly during times of peak labour demand that the peasants leave the farms, in order to go to town to market their produce.[31] Seen from a perspective of opportunity costs, the time that is thus diverted should obviously be given a considerably higher price than simply the inconvenience that is inflicted upon the peasants.

Considering that these problems derive from the ideologically motivated prohibition against private middlemen, it is fairly understandable that there is a certain reluctance to gather information regarding the costs involved. The ideological dimension, however, carries a certain importance of its own and we shall have reason to return to it at greater length below. Let us now proceed to the other aspect of the case against the private sector – that of speculation. Here we shall find the same type of negative attitudes, only derived from a different set of arguments.

The Problem of 'Speculation'

An interesting illustration of the arguments that are typically raised in relation to the problem of speculation can be found in a miniature debate on the activities of rural 'strawberry kings', which was published in a series of issues of *Komsomolskaya Pravda* during 1983. In the initial article, which sparked off the debate, the paper's correspondent captures much of the ambivalence contained in the general debate, as he first states that the country appreciates and supports the valuable contribution that is made by the private sector, and then goes on to make the following cautionary remark:

> At the same time, we must not close our eyes to the negative tendencies that can be found in the use of the household plot, to the fact that sometimes such activities are transformed into a main

> source of income, and thus give rise to relapses into petty-bourgeois psychology.[32]

This is an obvious illustration of that argument on 'too much' that was referred to in the above discussion about private versus personal. How much is fair? How large an income can be derived from the plot before its cultivator is crowned a 'strawberry' or a 'cucumber' king, before he is accused of being a 'speculator' who profiteers on the state and on his fellow citizens? This question certainly begs an answer.

Another peculiar aspect that is brought out by the same article concerns specialization. In the Soviet press one can regularly find accounts of private plots specializing in this or that high-value crop, be it citrus, tomatoes or, as in the case at hand, strawberries. It is certainly understandable that some people are upset about the profits that are derived from such activities, when carried out on a large scale, but it is perhaps a trifle peculiar to note that members of the gardening associations are formally prohibited from doing so:

> We may also note that it is not permitted to use the plot for growing only one, or maybe 2–3 products, including flowers in such amounts that clearly exceed the needs of the household. Nor may it be used for purposes of profits and enrichment.[33]

Another peculiar illustration of the ambivalent stance towards activities on the plots can be found in complaints that are voiced about the fact that those who specialize in this fashion have the audacity to purchase other foodstuffs in the state stores: '... even vegetables are bought in the stores, while the household plot is filled from one end to the other with strawberries.'[34]

In the West, few people would probably be upset by such behaviour, which theoretically at least ought to lead to an increased total output. For a Soviet citizen, however, there is the onerous problem of standing hours in line for scarce produce, which may explain that any extra pair of seemingly 'unnecessary' feet in the line will be ill received. Moreover, there is an additional and rather specifically Soviet problem, which was pointed out by a representative of the consumer cooperation Tsentrosoyuz in a subsequent article in that same paper: 'Some people specialize in one crop, sell it at high prices, and purchase the remainder of their needs at low state retail prices.'[35]

The real root of the latter problem is of course to be found in Soviet price policy, which fixes low prices for the official outlets, where little food is found, and allows high prices on the free (*kolkhoz*) markets, where a considerably wider selection is on offer. In a sense, we can even interpret this price policy as a precondition for the very existence of the private sector, as we know it today. As we shall see below, however, quite illegal attempts are regularly made at interfering with the free formation of prices on these markets, practices which must certainly be welcomed by both ideological watchdogs like Rutkevich, and by the urban population at large, but which also contribute an additional – and important – form of discrimination against the private sector.

It is also a well-known fact that much produce in various illicit ways ends up on the markets, thus giving rise to substantial 'unearned incomes' (*netrudovye dokhody*): 'Some citizens, who do not participate in socially useful labour, buy produce from the rural population and resell it in the cities at high prices,' admits the same Tsentrosoyuz official quoted above.[36] While this represents a rather mild form of private middleman activity, there are of course also cases where produce is illegally diverted from the state stores and on to the free market, a practice which at times can acquire quite substantial proportions, involving large sums of money as well as government and party officials in high positions.[37]

The various forms of activities that take place at the *kolkhoz* markets will be subjected to closer scrutiny later, but it may be useful at this point to note that the ideological struggle that is continuously waged against such trade is in some respects a rather unpleasant one, since in many cases no formal laws are broken. What is demanded from the peasants, in such cases, is that they should refrain from profitable – and quite legal – activities, simply because these would earn them more money than what is considered 'fair'. Exactly where the limit should be drawn has not been established, and from a legal, as well as moral, point of view, this is of course highly unsatisfactory.

The Legal Position

What then is the legal position regarding private agricultural activities? First and foremost, there is the prohibition against private middlemen, which dates back to the 1920s. During the period of the New Economic Policy (NEP), private traders – the infamous NEP-men – quickly put the official trading network to shame, by proving themselves capable of offering the peasants far better prices, as well as more reliable purchasing procedures. Ever since the end of NEP, such trade has consequently been illegal, being generally referred to and punished as 'speculation'.[38]

What *may* take place is expressed as follows, by Article 240 of the Civil Code of the RSFSR:

> The sale, by *kolkhozy*, of surpluses of agricultural produce, which have not been purchased by the state, as well as the sale by the citizens of produce belonging to them, may take place at prices which are agreed upon by the parties concerned.[39]

Three points are important to note here: first the free formation of prices, second the provision that you may only sell what 'belongs' to you, i.e. what has been produced on your own plot, and third the absence of any restriction of marketing rights to the rural population alone. The latter in particular is not always clear in Western writings on the subject.

In addition to the prohibition on trade, there are also the rules on labour participation in the socialized sector, which have been mentioned above and which make it illegal to run the plot as a full-time occupation, or even as a main source of income. Otherwise, however, we are left with the considerably more complicated issue of 'unearned incomes', which has long gone without any form of precise definition. In the Soviet Constitution, it is clearly stated – in Article 13 – that the basis of private activity must be income from 'work' (*trudovye dokhody*), but no precise definition is given.[40] In the spring of 1986, a decree was issued which signalled a veritable witchhunt on the sources and recipients of 'unearned incomes', but still without providing any precise definition of the target.[41] Then, in November 1986, a law was passed *allowing* certain specified 'individual' activities for private gain, but here all reference to the peasantry was conspicuous by its absence.[42]

The pronounced official ambivalence on this point is admirably captured in a recent *Krokodil* cartoon, which shows two militiamen watching an old peasant sitting in a corner making baskets on a *kolkhoz* market somewhere. One asks the other the following rather pertinent question: 'Should we fine him for unearned incomes, or encourage his individual labour activity?'[43] As far as private agricultural activities are concerned, we are thus still moving in a grey zone. This is a fact of crucial importance to bear in mind when we arrive at our discussion of the practical manifestations of those negative attitudes outlined above. The absence of clear rules and definitions obviously leaves considerable scope for randomness and abuse and this, as we shall see below, has serious implications for the formation of attitudes amongst the peasants.

So far, our presentation has been clearly slanted in the direction of presenting a picture of peasants victimized at the hands of ideologically narrow-minded bureaucrats, and – it must be admitted – this has not been done without a purpose. Since our main ambition in this study is to investigate the impact on the peasant producers, and on their urban 'fringe' colleagues, of discrimination and hostile official attitudes, such problems of semi-legal and outrightly illegal speculation that glare in the eyes of responsible officials have not yet found their deserved place in the discussion.

At this point, however, the time has come also to look at the reverse side of the coin. Since the next chapter of the study will be devoted to investigating various forms of discrimination – and at times outright harassment – against the private sector, it is necessary to gain some impression of at least some of the factors that contribute towards the formation of such negative attitudes. An impression of 'evil' officials would of course be both simplistic and unhelpful for the purposes at hand. Let us therefore take a somewhat closer look at what is hiding behind the frequent allegations of 'speculation'.

On and Beyond the Limits of the Law

The arguments in this section of the chapter shall be presented in declining order of illegality, starting with such clearly criminal activities which, if carried out on a large scale, would render long prison terms, proceeding via such formally illegal – but still widespread – private middleman activities, which would normally be dealt with by purely administrative means, and concluding with such formally legal activities which, in terms of what Soviet citizens consider to be 'right and proper', may well be broadly considered to be unfair or immoral.[44]

The first point on the list will be illustrated by a survey of corruption on the *kolkhoz* markets, which was made by *Sovetskaya Rossiya* in 1985.[45] The picture presented is rather distressing, as it is claimed that no less than 50 per cent of all officials at the markets eventually end up in court. In a recent example, twelve market directors in a major Russian city were arrested, one after the other, during a period of two years. Eventually their fate was shared by the chief official in charge of all of the city's markets. No doubt, this is to some extent a manifestation of that campaign against corruption which was once launched by Andropov. The trials, however, are of course also indicative of the inherent rot in the system.

The problem is that there is simply too much money to be made, for the temptation to be resistible. According to specialists consulted by the paper, some 65–70 per cent of all those trading at the markets in the Russian republic are big-time operators, with a *daily* turnover that ranges from a few hundred to several thousands of rubles. For the sake of comparison, it may be noted that the average *monthly* wage of an industrial worker is somewhere in the vicinity of 200 rubles. A few examples are provided in order to illustrate the temptations involved.

In one case, a person had bought 250 kg of citrus from a state store, at 2.50 rubles a kilo, and resold it at the market, at 6.00 rubles, pocketing a profit of 875 rubles, or more than four months' wages for a normal worker. In another case, half a ton of onions had been bought at 50 kopeks to the kilo and resold at 2 rubles, thus earning the trader a similar profit. Needless to say, one simply does not just walk into a Soviet food store to buy half a ton of oranges, so there is obviously a fair bit of profit-sharing going on as well. As for other costs, however, these largely come out of the public pocket. Another example will serve to illustrate.

Some five tons of apples are 'procured' from a state outlet and resold on the market, at four rubles to the kilo. The costs incurred by the 'trader' during the two weeks it took to unload the shipment amounted to six rubles per ton in storage costs, plus 31.40 in market fees, or a total of 61 rubles and 40 kopeks. The net return was 12,000 rubles. The fact that the trading fee is a flat rate, and not a percentage, is also underlined by an illustration. In this case, a man from one of the southern republics brought two tons of apples for sale, a quantity which hardly originated solely from his own plot. He proceeded to sell the apples at four rubles to the kilo, earning a total of 8,000 rubles. Nevertheless, he still had to pay the same fee as the *babushka* standing next to him, with no more than fifteen bucketfuls. The feelings expressed by the *babushka*, to the paper's correspondent, underlines that hostile attitudes towards the 'speculators' can be found not only among officials and the general public, but also among the 'little' traders. The latter will of course not only envy the large profits, but will also suffer unjustly from the general attitude against speculators.

Given the large amounts of money that are involved, it is hardly surprising that officials at the markets will be open for 'gifts' of various kinds, ranging from pure profit-sharing, in order to overlook forged freight documents, to smaller handouts for allocation of the better spots on the trading floor. In one case, of a raid against people selling flowers on a market, the militia managed to gather a total of some 80,000 rubles' worth of flowers, that were left behind by the fleeing trader-speculators.[46] The valuation, moreover, was made at the fixed state prices, and the mind boggles at the thought of what they would have brought on a day like 8 March – international women's day – when the price of flowers may run as high as 4–5 rubles apiece. It is hardly surprising that people like these will have some extra cash to spare for those responsible for law enforcement to look the other way.

A crucial point to note in this context, however, is that far from all of these big traders are actually operating on the wrong side of the law. Via a proper specialization in production, and a good knowledge of the conditions of demand and supply on various markets, quite respectable – and fully legal – profits can be made. Transport in the Soviet Union is cheap, and as we shall see below, many inputs to plot production (viz. land, water and power) are subsidized. This brings us to the second point on our list, that of the broad range of semi-legal activities.

It used to be the case, that any peasant who wanted to market produce from his plot would need a certificate (*udostovorenie*) from his *kolkhoz* or *sovkhoz* in order to prove that the stated quantity of produce had in fact originated on his own plot, and was not the result of any market transactions. These certificates were then checked by responsible officials, before trading could begin. In practice, however, it is obvious that the possibility of actually verifying that only the stated quantities will be sold is seriously limited, and today the demand for certificates seems to have been dropped:

> Yes, no matter how strange it may seem, such documents are no longer required – such is the present regulation of market trade. What it was that brought about such a principally important decision, nobody at the Mintorg [Ministry of Trade] has been able to explain in any sensible way.[47]

It is thus believed to be a common practice for some people to specialize in standing at the market, for friends and relatives in the villages or for simple profit. The number of familiar faces seen in the market stalls confirms this: 'For months in a row these southerners sit at the market. Do they really all have 3–4 month holidays? And back home there is a labour shortage. Others have to work double shifts.'[48] The point about labour shortage in the South is of course in poor correspondence with reality, but it illustrates the feelings that are involved. The activities in question are formally quite illegal and punishable by law on the two counts mentioned above – illegal trade and illegal absence from work in the socialized sector. If, however, they are sufficiently widespread, effective enforcement will obviously be a virtual impossibility.

If we stick to the example of flowers, which is of course especially sensitive since it implies diverting resources from food production, we can quote some interesting evidence from an article in *Komsomolskaya Pravda*, entitled 'Flowers of Evil'. The story is focused on a village where the production of flower seeds had crowded out 'normal' activities on the plots. More than 2,000 out of a total of 2,655 households in the village were engaged in seed production, earning in a season 'a few times the annual income of an engineer or a teacher.' The average monthly income of workers in the village was said to be less than 150 rubles. More than 600 able-bodied workers had completely ceased to work in the socialized sector. The comments made by the paper are of great relevance for our perspective, as it emphasizes how these profits have started to influence people's mentalities, to 'sow the impression that money can buy everything in the world.' The village is seen as evidence of how money has 'made enemies of the closest friends.'[49]

To be upset about those more or less criminal forms of 'speculation' that have been outlined above, is one thing. The real problems, however, start when attitudes that are derived from observing such activities are reflected against the private sector as a whole, and particularly against activities in the sphere of production. This brings us to the third point on our list, that concerning formally legal activities which are considered to be morally unjust. As an illustration, we shall use another miniature debate, this time from *Pravda*, which leads off with an account from a *raion* in the Ukraine that had experienced a veritable take-over by the private sector.[50]

The *raion* in question is presented as having excellent conditions for vegetable growing. In order to further improve the gifts of nature, the local peasants had undertaken considerable investments. According to a survey that had been carried out by a special commission, more than 5,000 greenhouses, with boilers and electric heaters, had been illegally constructed. For irrigation purposes 8,000 wells had been drilled, and water tanks with capacities ranging up to 6 m^3 constructed. Electric pumps were humming around the clock, and former cow sheds were filled with fertilizer, herbicides and other chemicals. Orchards had been cut down, vineyards uprooted and animals slaughtered.

Needless to say, returns were considerable. The local gardeners were reported to market some 60–80 million rubles' worth of early vegetables every year, representing a sum four times larger than the total wage fund for all those employed in the socialized sector of agriculture in the *raion*. At the same time, losses of labour in the socialized sector were estimated at 20–24 thousand man-days in the local *sovkhozy* alone.

It should thus come as no surprise that the author of the article – the local *raikom* secretary – shows obvious signs of being upset:

> What then is the solution? I may be wrong, but as far as I can see what we need today, on a nation-wide scale, is not only measures to promote personal support farming, but also such measures that can put an end to the spread of profiteering.

As counsel for the defence of the plots, we find Shmelev, who turns the matter 180 degrees around by arguing that what is needed is more support for the private sector. By supplying it with purebred animals, adequate feedstuffs and proper tools and implements, the already substantial contribution that is made by this sector towards the country's overall food supply could be yet further increased. The more precise nature of the required support is discussed in Chapter 5. We may note, however, the two diametrically opposite interpretations of the situation at hand.

In summing up the arguments presented, *Pravda's* correspondent captures some of the really crucial issues involved, by establishing quite clearly that the real explanation of the large profits that are thus derived should be sought not so much in terms of high productivity, as in the very special circumstances that surround plot production. First of all, a large amount of 'social labour' is allegedly diverted from the socialized sector, via the different forms of support that are given to the plots (e.g. young animals, feed, seed, machinery, etc.). In addition, land and various overheads, such as water and electricity, are provided at low or token cost only.[51] All of this is seen to produce 'unearned incomes', particularly so if we consider that a part of the output is sold at high prices on the *kolkhoz* market In the latter sense, the private sector is even accused of acting as a parasite on the socialized sector:

> Facts indicate that certain persons, some of whom are permanently employed in the socialized sector of the farms, are increasingly turning their efforts to developing the private sector, which is then losing its 'support' character. This has a harmful influence on the socialized sector, and on people associated with it.

What we have here is yet again that attitude of 'too much', which was illustrated in our above discussion of the 'strawberry kings'. Free or heavily subsidized inputs, in combination with high prices on output, can be tolerated as long as nobody uses 'too much' of these inputs, in order to produce – and market – 'too much' output. Yet, under a policy that is allegedly aimed at supporting the private sector, it is certainly legitimate to wonder where the logic has gone. Is the new policy aimed only at supporting the plots up to the point where they market, say, 8.5 per cent of their total output? If so, it would no doubt be a good move to allow all of those involved to know exactly where that limit is, in order to avoid accusations of the kind that we have seen on a number of occasions above, and which will reappear again below. What is involved here is quite simply the concept of a *Rechtsstaat*, the absence of which creates a number of serious problems. Let us summarize the points that have been raised in this chapter.

Conclusion

We started out above by looking at the important issue of 'private' versus 'personal' agriculture, and saw that there were indeed good reasons for some form of a distinction to be made in that context. While it is standard Soviet practice to do so for chiefly ideological reasons, however, our reasons were seen to be connected with various aspects of the process of *production*. Since the holders of private plots are critically dependent on the good will of the managers of the socialized sector, for a number of vital inputs, and since the latter depend on the former for labour effort, we are not really dealing here with two distinctly separate and alternative modes of production. Instead, the two sectors depend on each other in a rather intricate way, which provides fertile ground for the growth and spread of all those attitudes of hostility and suspicion that form the main theme of this book. The real importance of the issue of 'private' production has been sought in relation to the moral and ideological aspects of that concept.

The arguments that we have seen raised above, against the private sector, are all strongly influenced by problems connected with the existence of a 'private' sphere, inside the socialized sector. Seen from an ideological perspective, this is of course to be expected, but beneath the ideological surface we can also discern a set of arguments which are dressed in a more serious guise. On the one hand, we have arguments that the private sector acts as a brake on development, by diverting labour effort from the socialized sector and by fragmenting the use of other resources. On the other, we have such arguments that focus on 'speculation', and on 'unearned incomes' from various private activities.

Taken together, these two lines of argument bring out a clear case of 'second best', with unfortunate consequences for the development of the agricultural sector as a whole. Since the socialized sector is so poorly developed, the plots cannot be abolished. Given, however, the reluctance to allow 'unearned incomes' they cannot be permitted to develop into an efficient sector either, a sector which could have found a natural role within the system, according to the principles of specialization and comparative advantage. Instead they act to divert resources from the socialized sector. Consequently, the argument regarding the function of the plots as a 'brake' on development becomes self-fulfilling. A similar point was made above with respect to the *kolkhoz* markets. What we get is a process of cumulative causation, where ideological arguments against private initiative serve to stifle economic efficiency and thus find their own justification. This process is unfortunate not only in narrow economic terms, but perhaps even more so from the *Rechtsstaat* point of view that was referred to above, i.e. that none of those involved can be sure of what the *actual* rules of the game are.

Before concluding, we must also observe that the ideological 'case' against the private sector is further strengthened by the – well-publicized – occurrence of both outrightly criminal forms of 'speculation', and of such activities that are most likely considered immoral and unfair by broad sections of the Soviet citizenry. Both of these have been demonstrated above. Noting the latter point is of course not the same as an accusation against those who simply take advantage of given circumstances, but it may certainly help to explain some of the hostile attitudes that prevail amongst sections of both the bureaucracy and the population at large. To these we shall now turn.

Chapter five

The practice of harassment

Viewed against the background of the various negative attitudes that have been described above, in relation to the private sector and its inhabitants, it may perhaps be somewhat easier now to understand some of those discriminatory – and at times – outrightly illegal acts that are committed by various officials and authorities, chiefly at the local level, against those who are exercizing their constitutional rights to private agriculture. The purpose of our examination here of the nature and frequency of such actions is twofold. First of all, any type of discrimination against the private sector will obviously have a detrimental influence on its productivity, and will thus be costly in economic terms. The fact that such actions and attitudes still persist consequently requires an explanation. This can be provided in one of two possible ways.

Either they can be understood in terms of a conviction that the private sector is in the process of 'dying out', in which case the short-run costs from a further squeeze will be offset by long-run benefits from a more rapid strengthening of the socialized sector. As we have argued above, this may well have been a sincere – albeit perhaps mistaken – belief behind Khrushchev's attacks on the plots. Given the present policy of support, however, it would not seem to offer a feasible explanation for what is happening today. What we are left with, then, is an explanation which departs from priorities given to private agriculture in the system of material supply, and from the behaviour of local officials charged with the supervision of agriculture. The factors that determine these two patterns will be an important topic for the remainder of our discussion in this book.

Our second purpose is somewhat more complicated, in that it shall attempt to identify the constituent elements of such moral and psychological influences that the various forms of discrimination must have on the peasantry. In the present chapter, this ambition will be pursued only indirectly, by means of providing a number of smaller pieces for the puzzle. These will subsequently combine to form an important background for our discussion, in Chapter 6, of the various types of peasant response that can be observed, and, in Chapter 7, of the consequences that such responses may in turn have for the overall economic and political system of the Soviet Union.

Harassment, however, is a fairly strong word which of course is applicable only to some parts of the official attitude to private agricultural activities. In order to distinguish here between simple suspicion and grudge, on the one hand, and active harassment and outright hostility, on the other, our presentation below will be divided into two parts. We shall start by looking at the milder manifestations of the problem, as represented by the simple withholding of such support that has by now been called for in a number of official decrees. Then we will proceed to its more sinister side, which is related chiefly to the process of marketing produce from the plots. In the latter case, local officials will be presented as guilty of quite criminal procurement practices. In the conclusion of the chapter, we shall try to outline the causes and motivations that lie behind these different types of attitudes and actions.

Withholding Support

It is of course well nigh impossible to substantiate statistically that the officially requested support for the private sector is being deliberately diverted or withheld, but there is plenty of indirect evidence. The simple fact, for example, that the last decade has witnessed four major Central Committee decrees, with the usual aftermath of various political speeches and statements, all repeating the need for support, is in itself hard to interpret in any other way than as an admission that reality fails to live up to expectations. A substantial need for support is also clearly discernible. Shortly after the publication of the 1981 decree, for example, the following could be read in *Literaturnaya Gazeta*: 'An amazing document. On point after point it was like hearing the voice of the masses, on a *kolkhoz* meeting somewhere, or on some bench up against the wall of a peasant hut.' [1] This statement should be seen against the background of the 1977 decree, on support for the private plots, and the appearance in 1986 and 1987 of two further decrees calling for much the same.

In order to disentangle the various aspects of the problem of support, we shall proceed in five separate steps. As background, we shall start by looking at the general situation of agriculture, in the system of 'material and technical supply'. Then we shall review the rather illustrative case of providing mini-tractors for the private sector, and from there we will proceed to the broader problems of providing a range of such tools and implements that are suitable for use on the plots. Our fourth point will deal with the problems of obtaining feed for privately owned livestock, and in conclusion we shall look at the situation that encounters the peasants on the *kolkhoz* markets.

The Last Place in Line

Agriculture in general has a low priority in the Soviet economy. At first sight this statement may seem a bit hard to swallow, given our previously presented picture of a massive resource commitment. How is it possible to talk of a low priority when during Brezhnev's years in power, for example, agriculture's share of total investment in the Soviet economy increased from 22 to 28 percent, and with the 1982 Food Programme calling for that level to be sustained throughout the 1980s? Yet, it is only by focusing on the systematically low priority that is *actually* received by agriculture, in industrial production and in material supply, that we can explain how such a massive resource commitment can yield such poor results. Thus we also hope to justify our previous use of the metaphor of comparing Soviet agriculture to one of the Black Holes of astronomy.

It is a rather well-known fact that much of the machinery and equipment that is supplied to Soviet agriculture is of a design and quality that leaves quite a lot to be desired. Complaints about this appear with monotonous regularity in the Soviet press and a few isolated examples will serve as representative.

At the beginning of 1982, i.e. around the time of the introduction of the Food Programme, *Pravda* reported about the introduction of a new heavy-duty all-round tractor. The tractor was designed for use with 33 different kinds of implements, but only four of those were actually produced.[2] Later on in the same year, *Sotsialisticheskaya Industriya* reported that the combine harvesters Kolos and Niva, which are designed for an uninterrupted operation of about 300 hours during a season, in practice have an average operating time between break-downs of no more than 5.7 and 8.7 hours, respectively.[3] Towards the end of 1986, i.e. in the midst of *perestroika*, Gorbachev was interviewed on Moscow television, saying that the development of new machinery is still lagging, and that quality remains poor,[4] and from a conference held at about the same time at the Minselkhozmash, the ultimately responsible Ministry of Agricultural Machine-building, the following statement was reported: 'When machinery and equipment is put into use, massive defects are revealed in welded joints, assembly work, adjustments and painting.'[5]

The reasons underlying this sad state of affairs are certainly of many different kinds, and the notorious problems of a low priority for quality in industrial production obviously rank high here. It is interesting, however, that some authors find very specific reasons why agriculture in particular is being hurt. Yanov, for example, blames a 'militarization of agricultural machine-building'.[6] His argument is that many of the tractors that are used in Soviet agriculture are ill suited for such use simply because they are manufactured by enterprises that are under obligation to be able to switch into production for military needs at short notice.

Whatever the reasons, the observable consequences are that Soviet tractors are so heavy that they not only cause serious soil compaction, and thus reduced fertility, but also very easily get stuck in mud, which tends to be the normal state of many Soviet country roads in spring. In addition, they are so unwieldy that they cannot be driven by women, who then choose to leave the farms. This type of discrimination against the needs of agriculture extends into the fields of all of its 'partners', i.e. not only machine building, but also construction, fertilizer, transport, repairs, etc.[7]

In addition to these problems of design and quality in production, we have the problem of material supply, which is particularly troublesome with respect to spare parts. For decades now, the shortage of spares has been a target of vicious criticism, above all in the satirical magazine *Krokodil*,[8] but seemingly without much of a result. The consequences are that new deliveries of modern and expensive machinery are turned into scrap, as spare parts are removed in order to keep existing machinery in operation. According to a recent Soviet source, close to one out of every two new tractors that are delivered to agriculture is destined for this fate.[9] In addition, there are also accounts of thefts, from warehouses and from railway sidings, which testify to the seriousness of the problem.[10]

All of this provides at least part of an explanation as to how it has been possible for Soviet industry to turn out, for the past couple of decades, from 300,000 to 400,000 tractors every year, without a remotely corresponding increase in the stock. Poor maintenance and a lack of spare parts has produced scrapping rates of such a magnitude that no more than 10–15 percent of the new deliveries have been *net* additions. A similar situation can be identified with respect to combine harvesters and trucks.[11] In an interview with *Moscow News*, in early 1988, the highly respected Soviet agricultural economist and academician Vladimir Tikhonov made the following comment:

> Every year industry supplies agriculture with more than 400,000 tractors. For comparison's sake, we can say that 3.5 million farmers in the USA are supplied with not more than 80,000 tractors a year. The service life of a tractor in our economy is an average of 7–8 years. This is half as much as American, and at least a third less than the service life our tractors are designed for.[12]

Given this overall situation, and in the light of what has been said about official attitudes to the private sector, it is perhaps not hard to visualize that a mere trickle only of resources will eventually reach the private plots. What makes the situation particularly difficult for the latter, however, is the fact that even benign *kolkhoz* or *sovkhoz* managers are seriously limited in what they can do to help. The socialized sector's machinery and equipment are intended for large-scale field operations, and are consequently of limited use to the plots. Whatever help can be rendered is restricted to the accessible fringe of the plots, and to those distant corners of the collective fields that are placed at the disposal of the private sector for pasture and silage. It is certainly hard to visualize a heavy-duty tractor or combine harvester passing through the gate of a fence around a private plot, skirting some apple trees, and carefully avoiding the strawberries, in order to get to the potato patch.

What is really needed here, is entirely new types of machinery and implements, much along the lines of the intermediate technology that is often advocated for Third World agriculture, i.e. mini-tractors, hand cultivators, weeders, reapers, etc. Ironically, many of these are widespread in private gardens in the West, but few and far between in the Soviet Union. Although this lack of equipment is first and foremost a problem for the plots, it is of some considerable importance to note that needs are reportedly great in the socialized sector as well, i.e. in that very sector which is repeatedly called upon to render help and assistance to the plots.

One Soviet source claims that there is in the overall agricultural sector some 90 million hectares of agricultural land (about 15 per cent of the total) which is of such poor quality (marshes), of such small size (forest clearings), or in such difficult locations (hill sides), that the use of 'traditional' techniques of cultivation is not feasible – or even possible.[13] The introduction of intermediate technology would in such cases lead to considerable savings, of various kinds. Most importantly of course, it would reduce labour demand – according to our above source by some 15 per cent – but it would also allow economies on fuel and metal, as the new equipment would be of considerably smaller size than that which is currently in use. Moreover, it would allow the use of 'additional categories of workers', such as women, children and pensioners, who are normally not able to handle such unwieldy machinery as is presently in use. The latter would of course render further help in easing the labour bottleneck. In spite of all these potential benefits, however, 'we must acknowledge that the tractor and machine building industries have seriously lagged behind in solving the tasks at hand.'[14]

The broader range of such tools and implements that are needed will be investigated in greater detail below. In order to create a picture of the general difficulties involved, however, we shall start by taking a closer look at the specific case of the mini-tractor, a case which is fairly well publicized and probably considered to be representative of the whole complex of problems involved here.

The Fate of the Mini-tractor

The issue of supplying mini-tractors for the private plots surfaced for the first time in the Soviet Union back in 1967, and it is perhaps not surprising that the proposal came from the Baltic republics.[15] At the time, however, the situation was hardly ripe for such a bold move. As we have seen above, policy on the private sector after the fall of Khrushchev, in October 1964, was initially more that of restoring the previous order than of embarking on a new policy of active support. A decision to supply mini-tractors would in this context have been a clear move in the direction of support, and would thus have implied an important overall policy decision on agriculture.

Although the leadership might thus not have been quite prepared at the time to take such a policy decision, a cautious first step was taken in the following year, as the Minselkhozmash was charged with responsibility for 'material and technical supply' to the private sector. Funds were made available, a design office organized and there was even a factory set up, in Pavlovo on the Oka. In 1972, however, the Committee for Science and Technology decided to investigate what progress had been made and the findings were hardly encouraging.

As a result of this interest, an exhibition of tools and implements suitable for use on the plots was staged in Moscow in 1976. The only mini-tractor that was to be seen there, however, was the 'Rioni-2', a 200 kg, 6 horsepower unit, produced by the Kutaiskii Small Tractor Design Plant, which sold at the exorbitant price of 2,200 rubles. The combination of poor design and high price had completely ruled out all demand from the private sector, and the plant's entire output of some 2,000 units annually was instead absorbed by the construction industry. In a comment on the lessons to be drawn from this exhibition, *Izvestia* concluded that the tractor in question was a 'discredit to the entire idea of intermediate technology', and the paper chose to lodge the entire blame for the general lack of progress with the Minselkhozmash, claiming quite frankly that in practice that ministry had been using its allocated resources for purposes other than those intended.[16]

A couple of years later *Izvestia* is back with a second article on the subject, entitled 'What Happened to the Little Tractor?' After presenting its readers with an update on the matter, the paper presents a wealth of evidence from letters received, all of which indicate both a pressing demand and the fact that very little is actually happening on the supply side. Responsible enterprises and ministries are accused of being primarily interested in presenting excuses, focused chiefly on the lack of space and resources, and on the need to give priority to large-scale equipment destined for the socialized sector. The rather weak point of defence, that there would not be sufficient demand to warrant mass production, is singled out in particular, and the paper quotes a survey which had been carried out in Belorussia, indicating that 10,000–12,000 of these tractors could be sold within the first year of marketing. In conclusion, *Izvestia*'s reporter becomes quite indignant, arguing that the 'time has come to answer for this breach of government discipline', and that it is 'after all not a question of wishes, but of firm demands.'[17]

In 1980, however, *Pravda* publishes a letter from a farm director who claims that the demand for such tractors is still 'enormous', not only from the proper plots, but also from gardening associations and *podkhozy*. The director acknowledges that production has begun, but also points out that the current output, of no more than some hundred-odd tractors per year for the country as a whole, will in no way help solve the problem. The government is again called upon to take firm measures.[18] Some weeks later, the paper publishes a number of letters received that emphasize the lack of progress. Needs are still seen to be great, but continue to be met chiefly via various home-made arrangements. Prototypes exist, but there is no mass production. One author argues that mini-tractors should be explicitly included in the plans of the responsible Ministries.[19]

In 1981, it was announced that mini-tractors had actually been put into production,[20] but when *Izvestia* turns again to investigate the matter, in January 1982, the paper only manages to turn up two prototypes, one being the 'Rioni-2', and the other a model produced by the Minsk Tractor Plant.[21] The latter – known as MTZ-05 – is a considerably more versatile and attractive design, of 140 kg and 5 horsepower. Its intended attachments include ploughs, harrows, and cultivators, and it is capable of pulling a cart with a payload of up to 500 kg. In addition to the two models mentioned by *Izvestia*, there was also the 'Rosinka', a less attractive 1 horsepower model produced by the Kaluga Engine Works.[22]

All in all, however, while prototypes no doubt do exist, reports of large-scale production being initiated seem to lack foundation in reality. Instead we find numerous accounts of how people continue bulding their own makeshift contraptions, an undertaking which may produce outcomes of very different kinds. In some cases, we are told that people get into trouble for doing so, while in others, we may even find reports of peasants taking on contract work for the socialized sector, to be performed with their own little tractors.[23]

While the actual situation thus continues to look bleak, we must recognize that the official pressure continues to mount, no doubt as a result of the generally favourable official policy towards the plots. Shmelev, for example, reports that an annual output of some 15,000–20,000 units of the Minsk MTZ-05 mini-tractor is envisaged for 1986–90.[24] It remains to be seen, of course, what will come out of this, but on past experience, expectations should certainly not be overly optimistic.

Other Tools and Implements

Those difficulties that have characterized the history of the mini-tractor can, as we have indicated above, be observed in relation to the whole range of tools and implements that are needed on the various kinds of plots. In order to create a better understanding of the more precise needs, an official study was commissioned back in 1977, the result of which was a nomenclature (*inventar*) of 313 different kinds of equipment. At the time, the production of these tools and implements involved no less than 36 ministries and state administrations, supervising together some 255 factories, but the results of their combined efforts were not very encouraging.

In the 1977 *Izvestia* article quoted above, in relation to the mini-tractor, the paper also approached the broader issue of tools and implements, concluding that only four of the 313 items on the nomenclature in question were both new and useful, the others being either completely outdated or in need of serious improvements. The causes that were seen to lie behind this deplorable situation are similar to those outlined above. Factories view products designed for the private sector not as a second but as a *third* range priority, they use poor quality inputs, and their products in many cases last no more than one season.[25]

The consequences of this neglect are fully appreciated by the paper, as it is claimed that an adequate supply of good-quality tools and implements would result in massive savings of labour in the private sector. Potato production is singled out in particular, with potential savings of some 840 million man-days annually, but other crops would also benefit, thus bringing the total saving up to an estimated 1.5–2.0 *billion* man-days annually, a figure which is said to correspond to roughly *half* the total labour time expended in the sector as a whole.[26] No sophisticated marginal economic analysis is necessary to understand the implications, and *Izvestia*'s reporter is obviously upset in drawing the following conclusion:

> It is time – it has long been time – to deal with this problem, as they say to 'roll up the sleeves', to organize the production of suitable, reliable and relatively inexpensive tools.

By 1981, however, according to Shmelev only 140 of the items on the nomenclature were produced, and only the simpler kinds at that.[27] The actual situation, moreover, is worse than even this sombre picture would indicate, since there is an important difference between a piece of equipment being produced in small amounts by a single factory somewhere, and its general availability. According to figures from the consumer cooperation Tsentrosoyuz, which is responsible for providing the rural population with tools and implements, its annual sales amount, *inter alia*, to some 320,000 separators, 250,000 chaff-cutters, 40,000 churning machines and 4,000 shearing machines.[28] Considering the many millions of plots, and particularly the rapid expansion of the private fringe, this output must be seen as quite insignificant, *even if* the equipment in question had been of good quality, with an expected long life. The latter, of course, is normally very far from the case.

According to a survey made by *Sovetskaya Torgovlya* in 1982, there was in the country as a whole some 500 factories, supervised by 72 ministries and state administrations, which were devoted to the production of tools and implements, suitable for use on the plots.[29] The total output from these factories, however, was quite insufficient to satisfy demand. According to a subsequent article in that same paper, there was a shortage of *inter alia* some 1.2 million watering cans, 250,000 hotbeds and 400,000 sprayers.[30] A letter published in *Izvestia* some time later indicated a shortage of some three million shovels and more than a million scythes.[31] Shmelev sums up the problem in a rather conclusive manner:

> There is a great need not only to get the production of intermediate technology off the ground, but also to increase the output, and improve the quality, of simple tools used in the LPKh – scythes, shovels, rakes, hoes, pitch-forks, garden knives, hoses and pumps for irrigation, as well as other tools and implements, without which it is impossible to cultivate a plot or to raise animals.[32]

What we have here, is a rather clear illustration of the problem of industrial priorities, that was referred to at the beginning of this section. The poor status of agricultural development in general, and of support to the plots in particular, constitutes a major obstacle in the way of the current wager on the private sector. Perhaps most serious, in this context, is that the main problem is primarily one of quality, rather than of quantity. The flood of complaints about the poor quality of farm implements that is regularly published in the Soviet press presents a rather distressing picture in this respect.

The author of one such letter has the following to say: 'I bought two shovels, went into the garden, and started to dig around the currant bushes. I pressed the shovel twice with my foot, and it bent like a piece of cardboard.'[33] In another case we are told the following, about the quality of scythes: 'They must be sharpened every metre, and you must wield them with all your strength in order to get any results.'[34] A third author asks the following, slightly sarcastic question: 'Surely it cannot be impossible to make watering cans from which water would run out only at the spout?'[35]

The list of similar – and even more indignant – complaints could be made much longer,[36] but it is probably a more important task to identify the underlying causes of the problem. Here is Shmelev again:

> Such enterprises, however, that produce tools and implements for the rural population do so as a secondary line of production only. And from this stem the results – frequently poor quality of the products, an incomplete assortment, and an unsatisfied demand for even the simplest farm implements. Would it perhaps not be more efficient to concentrate the production of tools necessary for the LPKh in some large specialized factories?[37]

Again we have an illustration of the problem of industrial priorities, where the private sector is systematically placed at the end of the line. As Shmelev indicates, the production of tools and implements for the plots is used as a buffer, to absorb inevitable fluctuations in production. If there is spare capacity – fine. Otherwise the side-line will be closed, in order to guarantee a steady output of the more important lines of production. Against this background, the suggested concentration of production in specialized factories to meet the needs of the private sector would mean a dramatic improvement. This would also provide a litmus test of the true determination behind the policy of support for the private sector. It is hard indeed to imagine that happening.

The seriousness of the situation is further illustrated by the fact that some sources have even been suggesting that the horse, the old pivot of peasant agriculture, should be reintroduced. A long article in 1981 in *Komsomolskaya Pravda*, for example, quite frankly termed the horse a 'part of the cultural tradition of the rural population', and bemoaned the fact that while 1916 Russia had a total of 38 million horses, in the Soviet Union of 1981 only 5 million could be found. As a particular anachronism, the paper presented the fact that one may 'buy a 90 horsepower automobile, but not a natural one horsepower on four hooves.'[38]

As a reflection of the wide response from its readers, the paper returned to that same question in a subsequent issue, publishing a number of letters which placed particular emphasis on the importance to the private sector of having access to horses, but also pointed out that the use of heavy tractors in the socialized sector in many cases amounts to 'shooting birds with a cannon'. The fact was recognized, that ownership of a horse might be uneconomical to the single household. The author of one letter, however, suggested that maybe a number of families with private plots, say five to ten, could own a horse jointly. A call was issued for alterations in legislation on this point.[39] The thought of a renaissance for the horse seems attractive to Shmelev as well: 'Would it not be incorrect to delete completely from the inventory of intermediate technology that piece of equipment of one horsepower – the horse itself?'[40]

The simple fact that it has not been permitted by the 1969 *kolkhoz* charter to keep draught animals on the plots, is of course a formality of the kind that can be dodged or removed and it is rather interesting to note that this is precisely what has happened quite recently. As we have mentioned above, the 1988 *kolkhoz* charter explicitly allows peasants to keep 'productive' animals. Given its great versatility, the horse would of course be a great asset to the private sector. It would not only add important draught power, for transport as well as fieldwork, but would also provide an additional source of organic fertilizer. Given the long-standing difficulties in obtaining mineral fertilizers, the latter would certainly be of value.

There are, however, serious obstacles in the way of such a 'progressive' innovation, obstacles that are of a quite tangible nature. Where, for example, should the new horses come from? In accordance with Lenin's classic slogan that if the peasants were provided with 100,000 tractors they would turn communist, Soviet agriculture has been largely freed of its workhorses.[41] It is certainly difficult to envisage the unfolding of a vigorous programme of breeding new horses, destined for the private sector. Where, moreover, should such horses be kept? No suitable buildings will be found on the plots, so either private stables would have to be constructed, or some collective arrangement sought, neither of which seems very likely to occur. How, finally, should the private sector be provided with feed, which is already in desperately short supply for existing animals? The latter question is perhaps the most troublesome one. This brings us on to the fourth point in our list of problems in relation to support for the private sector.

Problems of Feed Supply

If we look at the *structure* of costs of production on the private plots, two features stand out: on the one hand the crucial role that is played by feedstuffs, and on the other the insignificant one that is played by tools and implements. In the years 1976–80, for example, feed accounted for 70 per cent of total costs (excluding labour), transport for 16.5 per cent and seeds for 4.5 per cent. Fertilizer was given as 'hardly measurable' whereas tools and implements were not even mentioned.[42] Furthermore, if we examine the sources for obtaining feedstuffs, we find a clear illustration of that dependence on the socialized sector which has been repeatedly referred to above. During the years 1976–80, only about a third of the feed used for private livestock was actually produced in the private sector, whereas close to half was received from the socialized sector (15 per cent as payments in kind, for labour performed, 11 per cent in the form of pasture on the collective fields, and 19 per cent according to contracts for the fattening of animals). In addition, 18 per cent was said to originate out of 'other sources'.[43] The latter includes thefts, as well as the long-standing – illegal – practice of purchasing bread in state stores, to be used for feed, a practice which is particularly important to the non-rural part of the private sector, i.e. the private fringe.

If the overall supply of feedstuffs had been adequate, this dependence on the socialized sector would not necessarily have been a great disadvantage, but as we know from our previous discussion feed shortage is a problem for the agricultural sector as a whole, and a main cause of Soviet grain imports. In this situation it will obviously be relatively easy for local farm managers, with a grudging attitude towards the private sector, to withhold feed resources. Hence, it is probably fair to say that problems in this respect presently constitute *the* major obstacle to expanding private sector production.

According to data presented by Shmelev, the private sector's share of total feed resources used – measured in terms of feed units – has declined steadily, from 30.7 per cent in 1965 to 23.2 per cent in 1983. Measured in absolute terms, this decline reflects the fact that while the socialized sector increased its feed use by 52 per cent, the private sector registered an increase of merely 15 per cent. What Shmelev does not mention, is that over the period as a whole a fairly similar decline was registered in the private sector's share of total livestock holdings.[44] This certainly reduces the impression of severity. If, however, we look only at the period after the 1977 decree on support for the plots, a somewhat different picture will emerge. During this time, private livestock holdings stabilized but the socialized sector expanded its feed use at almost twice the rate of the private (15 versus 8 per cent over the years 1975–83).[45]

One important indication that the feed problem is indeed a serious one, is the fact that animal husbandry is set in a process of losing its traditional role as the mainstay of private sector production. Today, only two out of every three families having private plots are reported to have any livestock at all, only one out of two have a cow, and less than half have hogs.[46]

Another important manifestation of the problems facing animal husbandry in the private sector is that of low productivity. The simple fact, for example, that the feeding of scarce feedstuffs to animals with minimal milk yields and weight gains seems like a rather unprofitable affair is also indicated by Shmelev, who points to the need to introduce purebred stock on the plots.[47] To do so, however, would require not only the provision of new stock, but also a true desire from the peasants' side to make the change. The latter is seen by Shmelev to be hampered by feed problems. As purebred stock is normally larger and heavier, it also requires more feed, and would thus only serve to exacerbate an already difficult problem.

It is of course officially recognized that feed supply is a major problem. In the 1982 Food Programme, for example, provisions were included for making payments in kind, not only to those actually employed in agriculture, but also to people on temporary attachments to help out with the harvest.[48] The importance of such payments is also illustrated by an interview with a *kolkhoz* chairman, published in *Izvestia* in 1982:

> According to my opinion, the problems of poor labour discipline started to be seriously felt when payments in kind were reduced, or to put it more specifically, when the norms for such payments were reduced. The guaranteed money wage is one of those measures the value of which it is hard to overestimate. Yet, we must consider also the psychology of the peasant.[49]

The latter point, about the effects of monetization of Soviet agriculture, has broader implications for incentives to work. We will return to this discussion below, but for the moment we shall dwell on the more narrow incentive effect of payments in kind. In a further example, from the above article in *Izvestia*, a *sovkhoz* director recalls how his farm had failed, after a poor harvest, to provide the agreed shares of payments in kind to those who had come from outside to help with the harvest. In the following year, many of the previous helpers did not return, although the *money* wages paid were reportedly good. The general shortage of feed for private livestock and the rapidly growing inflationary overhang have combined to reduce the value of money payments, and to increase the demands for payments in kind.

Let us proceed now to look at the *kolkhoz* markets, the last point on our list, and also the real *sine qua non* of the entire debate about real as well as alleged speculation.

The Kolkhoz Market

The *kolkhoz* market is – officially – the only sphere of the Soviet economy where prices are formed via the interaction of supply and demand.[50] Although it happens rather frequently that local officials intervene to control price formation, as we have mentioned above they have no legal rights to do so.[51] Formally, the concept of a *kolkhoz* market (*kolkhoznyi rynok*) refers to specially designated parcels of land, chiefly in the larger cities, where a legal trade in foodstuffs may take place. In practice, however, this 'market' also includes a large number of other places where trade takes place more or less spontaneously – such as railway stations, ferry landings, resort areas, street corners, and other places where large numbers of people are likely to converge. As Shmelev has indicated, the extent of trade on this fringe is not known, but believed to be substantial:

> What is sold there, and in what quantities, is unknown even to the Central Statistical Authority. We the consumers, however, know that it is a lot, especially in times of harvesting gardens and orchards.[52]

Historically, the *kolkhoz* market dates back to 1932, when a decree was passed legalizing such trade. As in the case of the private sector as a whole, this initial decision was a clear concession, forced upon the leadership by the famine that followed in the steps of forced collectivization. Political and ideological considerations simply had to be jettisoned, in order to promote the production of food, in any way possible. Since then, official policy towards the markets has been characterized by the same zig-zag course which we have observed above in relation to the plots.

Over time, however, the official markets have become permanent features of the Soviet townscape and food distribution system. Government officals are charged with veterinary and other health inspection, as well as with verifying that no unauthorized trade takes place. Peasants are charged a fee for trading there, and in some cases certain additional services, such as the use of cold storage or transport, may also be offered for a fee. In 1970, there was a total of 7,522 *kolkhoz* markets. By 1980, that figure had dropped to 5,862, but by 1986, a small recovery had taken place as a total of 6,092 was registered. The total number of trading spots on these markets in the latter year was close to 1.5 million.[53] No doubt, this latter development is a reflection of the change in official attitude towards the overall private sector, and we may perhaps expect a continued pressure on local officials to allow still further markets to be opened.

Although, as we have just indicated, the markets obviously play an important role in current Soviet society, in retrospect their importance has declined strongly. According to official figures, their share of the total turnover of foodstuffs in 1950 was 28.7 per cent, measured for a comparable basket of produce sold and at the actual prices paid. By 1965, that figure had dropped to 10.0 per cent. In 1975 it was 8.1 per cent and in 1980 it stood at 9.8 per cent.[54] In the 1980s, finally, the trend has stabilized, or perhaps declined somewhat, as the markets accounted in 1986 for 9.1 per cent of the total.[55] This impression of decline, however, is to some extent a false one, as *absolute* turnover has been rising steadily, from 3.6 billion rubles in 1965, to 4.2 billion in 1970 and 7.5 billion in 1980, a development which represents more than a doubling in 15 years.[56] Hence, an analysis of the importance of the markets in terms of income generated for the sellers provides a more impressive picture than if we simply examine it in terms of turnover.

This, however, is the picture that is provided by official Soviet statistics, and it is commonly accepted that it is a rather unsatisfactory one. Interviews with Soviet emigrés, for example, have indicated that considerably more is bought on these markets than is reflected in official statistics. Gur Ofer and Aron Vinokur have come to the 'temporary conclusion' that figures published for the Soviet Union as a whole 'are very likely understated by at least one-third and possibly one-half of their true value,'[57] and others have argued that even this represents an underestimate.[58] Based on a detailed scrutiny of the methodology used in producing the official data, Stephen Shenfield arrives at the conclusion that a corrective factor in the range 2.1–2.7 should be applied.[59]

This is significantly higher than the range of 1.5–2.0 which is implied by Ofer and Vinokur. If we apply Shenfield's correction to the above figures, we find that the *kolkhoz* markets account today for 10.5–13.5 per cent of the total turnover in foodstuffs, or for 19.1–24.6 per cent if measured according to a comparable basket of produce sold. Both are measured at the actual prices paid.[60] This, we may note, still excludes intra-rural *kolkhoz* trade, of which we know next to nothing. It is certainly easy to agree with the view that the quality of the official data, which really ought to be highly embarassing to the Soviets, is a derivative of an ideological reluctance to show the private sector for what it really is.[61]

The figures given above have all been measured in terms of the actual prices paid. There is, however, also another set of official figures which are measured in corresponding state retail prices. In 1980, the share of the free market in the total turnover of (comparable) foodstuffs amounted to 10.0 per cent according to the former definition, but to merely 4.7 per cent if measured according to the latter. The corresponding figures for 1986 are 9.1 and 4.2 per cent.[62] This distinction is of particular importance, in that it brings out two points of great relevance for our argument. While the former figure shows that the market still plays an important overall role in Soviet food supply, particularly in relation to perishables, which account for the really high prices, the *difference* between the two provides an important indication of the great differentials that exist between the fixed state retail prices and those prices that are paid on the free market. Let us start by loooking at the latter problem.

In the present case, we have a difference of more than 100 per cent, between the share of turnover measured in actual prices and that measured in state retail prices. This difference is the result of a substantial price inflation on the markets during primarily the 1970s. According to official data, the average price level on the markets in 1986 was 86.8 per cent higher than in 1970 and 13.8 per cent higher than in 1980. The main 'culprits' here are potatoes, fruit and vegetables, which recorded price increases of 136–146 per cent over the period as a whole, while meat and milk increased by a more modest 61–63 per cent and the prices of eggs actually fell. Seen in relative terms, market prices in 1970 were on average 68 per cent higher than state retail prices. By 1986 that figure had grown to 163 per cent.[63]

Substantial as these figures may appear to be, they are still only an average measure, which fails to reflect the fact that market prices vary strongly with the seasons. During early spring, for example, when fruits and vegetables are in extremely short supply, the premium that has to be paid on the market can rise to quite respectable proportions, and it is hardly surprising that the 'ordinary' citizens will be upset. Again, moreover, there are strong grounds to suspect that actual prices paid may be higher than those reported. Vladimir Treml, for example, has argued that official attempts during the 1970s to impose maximum prices have produced systematic underestimations of actual prices.[64]

Be this as it may, it still remains fairly obvious that the official attitude to the markets will be strongly influenced by the high prices. As we have seen above, those people who trade there are often quite broadly referred to as 'speculators', an accusation against which the proponents of the private sector predictably rise up in arms. Shmelev, for example, puts up the following defence:

> We must not agree with those who, without justification, characterize incomes derived from trade on the markets as 'unearned', since behind such incomes there lies work in the personal auxiliary sector (hard work at that) as well as difficulties in taking the produce to market.[65]

In addition to this reflection on the moral rights of being rewarded for hard work, Shmelev also – correctly –points out that 'speculation is a concept which is clearly defined by our Penal Code, as purchasing and reselling with the sole aim of making a profit'.[66] In cases where people are themselves selling produce from their own gardens, this paragraph is thus not applicable. As we have seen above, however, the problem is that precisely such speculation that *is* punishable by law is a rather common occurrence. The position of those who trade honestly on the market is obviously complicated by the existence of such activities. Even in their complete absence, however, we would still be left with the problem of 'unearned incomes'. Even if no *formal* laws whatsoever were broken, we would still have those two lines of ideological opposition that were presented above, i.e. that the markets – and the private sector in general – act as a brake on development, and that profits derived from such trade – although quite legal – are simply too high to be acceptable from a moral point of view. Even a formally honest trader will thus find himself in a 'no-win' position.

If we turn now to look at the markets from the important perspective of national food supply, we shall find a further illustration of that basic contradiction which we have seen above with regard to the plots, i.e. that on the one hand they play an important role, while on the other, they receive a systematically poor support. A fitting characteristic of this policy would be that of trying to get 'something for nothing', i.e. a desire that the private sector should increase its contribution without getting any more resources and without making any more profits.

The 1977 decree on support for the private plots was accompanied in the same year by a decree calling for support and help in developing the markets.[67] As in the case of the plots, the need for such support was evidently substantial. In the ideal case, the markets are housed in covered buildings, equipped with various sanitary and storage facilities and sometimes even have overnight facilities for the traders. Reality, however, is normally very far removed from this attractive picture. In 1979, Shmelev reported that only 4 per cent of all markets were covered, that only about a third had cold storage facilities, and that only one market in seventeen had a hotel or a hostel of sorts. Most importantly, however, only 69 per cent of the total *planned* investment in market improvement was actually realized.[68] The latter clearly reflects the problem of attitudes and priorities in the system of material supply.

In 1981, the year that witnessed the second decree on support for the plots, *Literaturnaya Gazeta* interviewed the chief health inspector of the Ukrainian republic. The picture provided by him indicated that the outcomes of the two 1977 decrees – on the plots and on the markets – show great similarities, in terms of a distinct lack of progress. A survey carried out in Kiev, by the state health inspectorate, showed that 11 out of the city's 19 markets failed to meet established health and sanitation norms. There was a general shortage of hot water, cleaning was inadequate and often undertaken only once a week, fruits and vegetables were rinsed in buckets that were also used for washing the floors, and in some cases there was not even a place for the traders to wash their hands. Investigations in other cities of the Ukraine had provided 'analogous' results.[69] Interestingly, the interviewed health inspector also used the occasion to stand up in defence of the 'speculators', by shifting some of the blame for the high prices: '[I]s it not our own carelessness and mismanagement that determines what prices will arise on the market?'

In the 1987 edition of *Narodnoe khozyaistvo*, finally, we are told that only 7.6 per cent of the markets are covered, and it may perhaps be warranted to quote in full the following table footnote on their general status:

> About 28 per cent of all markets lack electric light, and more than half running water. Less than 30 per cent of the markets are equipped with storage facilities, and less than 10 per cent have cold storage and freezer facilities. A hotel or a *kolkhoz* hostel can on average be found in seven markets out of a hundred, and only one in three has a restaurant of sorts.[70]

With this we shall conclude our discussion of the various aspects of inadequate support for the private sector, and proceed to the more sinister side of the problem – that of the more practical manifestations of such negative attitudes that have been dealt with above.

Active Harassment

As was indicated at the outset of this chapter, reference here to 'harassment' relates chiefly to such activities that serve to prevent the peasants from taking their produce to market. It should be noted right away, however, that our use of the word 'harassment' does not necessarily indicate any malicious intent on the part of local officials. As we may recall from our previous discussion, there are numerous cases of quite criminal activities which provide good reason indeed for taking action against speculation and corruption. The problem of harassment may thus to some extent be seen more as one of casting the nets too wide, than of the action being ill conceived *per se*.

In the present chapter we shall continue this 'defence' of abuse and malpractice, by recognizing that there may be additional cases, where local officials find themselves in situations where 'harassment' against the peasants is seen to be the only way for them to avoid getting into trouble with *their* superiors. In a recent study of Party involve- ment in agricultural affairs, Cynthia Kaplan has focused on precisely this problem. By presenting in great detail the very special (in relation to industry) conditions that have marked the role of agriculture in Soviet development, she has demonstrated quite conclusively how that 'petty tutelage' of local Party officials over the minutiae of day-to-day farm activities, which is so frequently condemned by Soviet media, is actually a quite logical consequence of the rules of the game rather than a manifestation of malpractice in some form.[71] In this perspective, the blame for abuse will of course rest primarily with the system as such, rather than with its individual officials.

Whatever the reasons may be, from the point of view of the peasants it is of course the outcome only that counts. In order, however, to provide a better understanding of the problem as a whole, we shall have to look at it from both sides, i.e. to include the rules facing the officials as well as the outcome that is inflicted on the peasants. To do so, we shall start by taking a very brief look at the performance of the official system of procurement and distribution of foodstuffs, which *ought* to fill many of those functions that are presently taken over by the 'speculators'. Shortcomings in this sphere create that dependence on the private sector which we have seen above to be accepted only with great reluctance.

After looking at this side of the problem, we shall proceed to review some illustrative examples from the Soviet press, of what may happen when the peasants are subjected to those various forms of illegal procurement practices that – from their point of view – will no doubt be understood precisely as harassment. While the former is solely intended as background, it is in the latter that we shall find the real 'meat' to go into our subsequent discussion of peasant reactions to abuse and malpractice.

The Official Traders

The main reason why the markets are allowed at all, in spite of all the allegedly negative moral and economic influences that have been outlined above, is of course precisely the same reason as that which accounts for allowing the private sector as a whole, i.e. that the state and cooperative networks are already struggling, trying to cope with their present responsibilities. Although much of the considerable volume of output that is produced in the private sector is consumed in the sector itself – by man as well as beast – its share in the overall marketed output of Soviet agriculture remains respectable, accounting in 1983 for no less than 11 per cent of the total.[72]

In all fairness, it should be observed that a large portion of this marketed production *is* handled by the official system, via direct purchases as well as via the increasingly important practice of long-term contracting between individual peasants, on the one hand, and various procurement agencies, on the other. Of particular importance on the latter count is a 1981 decree which permitted *kolkhoz* households to exceed established limits on livestock holdings, provided that they contracted to sell meat and milk to the socialized sector. This extra output could then be delivered by the *kolkhoz* to the state and be credited towards fulfilment of its compulsory procurement quota.[73] To estimate exactly how much of the private output that is marketed via different channels is notoriously difficult, particularly given our poor knowledge of intra-farm and intra-rural sales. We shall consequently not venture into any detail on this point.[74] We may note, however, that according to one Soviet source, state procurements accounted in 1982 for some 20 per cent of the meat, 47 per cent of the potatoes, vegetables and eggs, and as much as 67 per cent of the milk that was marketed by the private sector.[75]

Respectable as this may seem, it still leaves a large volume of produce to be sold on the market, which then becomes a grudgingly accepted necessity. Part of the reason why the official system is not capable of shouldering this additional burden is connected with an inadequately developed network of procurement points and with shortages of storage and transport. Refrigerated facilities are a particular problem here. Most importantly, however, we should recognize a number of features in the incentive system, which serve to discriminate against the private sector.

An important part of the responsibility for purchasing produce from the plots lies with the consumer cooperation, the Tsentrosoyuz, which is conspicuous by its almost total lack of interest in performing such services. In an article about the importance of trade to the private agricultural sector, V. Voronin gives the following three reasons for the poor performance of the cooperative network.

First, its overall responsibilities include a broad range of activities, where the system of plan indicators and bonuses is devised in such a way as to place purchases from the plots in a cleary inferior position. Second, although the authorities have long tried to pressure the cooperative into expanding the volume of its commission trade, i.e. to engage in the sale of private produce on a commission basis rather than focus solely on straight purchases, total turnover has remained a dominant plan target, which means that shortcomings in commission trade can be covered via a better performance in other activities. Consequently, only lip-service will have to be paid to the new directives. Third, and most importantly, since the share of perishables is considerably higher in the private sector than in the socialized sector of *kolkhozy* and *sovkhozy*, and since the present system of penalties and rewards makes underfulfilment of the plan for private purchases preferable to incurring waste and losses, purchases from the private sector will systematically be seen as being more costly. Consequently, it is hardly surprising to find examples where the Tsentrosoyuz flatly refuses to handle private produce.[76]

It is symptomatic of the short-sighted nature of Soviet incentives that in the latter case the immediate costs of produce going to waste will be passed on from the cooperative to the peasant producers, with little heed paid to the consequences for the system as a whole. Although this problem has long been the target of sharp criticism, every year large quantities of scarce vegetables continue to go to waste, or in the best case to end up as animal feed. The magnitude of the problem will obviously depend on how difficult it is for the peasants to make it to a market on their own, but in some regions as much as 30–40 per cent of the entire vegetable harvest may eventually end up as fodder.[77] That the problem is a persistent one has been indicated by Gorbachev himself, who claimed in his speech to the 27th Party Congress that total losses may amount to as much as 20 per cent, and that for certain products even 30 per cent may be reached.[78] Nikiforov is quite explicit about where he wants to place the blame:

> In the development of the LPKh an important role is played by the consumer cooperation. As studies have shown, however, this organization does not always dispose of its purchases by way of sales. The consumer cooperation may purchase, say, potatoes from the LPKh at relatively high prices, and then that produce goes to waste, is ruined, in those very same silos as that which is produced in *kolkhozy* and *sovkhozy*.[79]

An intuitive response might be to allow the market to step in where the state system cannot cope, but as we know from our previous discussion, the Party's experience of private traders during the NEP period was embarassing to say the least. Most probably, it still looms large in the minds of Soviet bureaucrats and ideologists. Private middleman activities are not only illegal, but may even over time have come to be considered as immoral by a majority of the Soviet population.

Short of such a drastic retreat, there is of course the possibility of abolishing the Stalinist practice of removing all produce from the farms at once, to be stored in centralized silos far away. There is certainly no longer any risk that the peasants would return at night and steal back what had been procured during the day, as was frequently the case during collectivization. The only obstacle against such a change in policy would seem to be simple inertia in making decisions about resource allocations. The strength of the latter should of course not be underestimated. As Nikiforov puts it, 'in these questions, it is necessary to take a stand against stereotypes, and to allow a variety of different forms of economic relations.'[80]

While such a change in policy on procurements may take a long time, one seemingly acceptable compromise already exists in the form of those so-called trade bureaus (*buro torgovlya*) which have been set up on some of the *kolkhoz* markets.[81] These bureaus accept produce for sale on a commission basis or, in cases where funds are available, may even purchase produce for resale. They thus serve to relieve the peasants of having to spend time selling their produce, in addition to the time that is already lost in taking it to the market. In spite of such potential gains, however, the system is very poorly developed. According to Shmelev, in 1978 there were no more than 717 such bureaus, in a total of some 6,500 markets. Moreover, they could not be found on *any* of the 28 markets in Moscow.[82] This neglect should be seen against the background of estimates from 1979, which indicate savings of some hundreds of thousands of man-days, at peak harvest time, that resulted from the work of those few bureaus that did exist.[83] The explanation is undoubtedly once again to be found in terms of inertia, expressed on this occasion as a successful defence from the side of the official supply system of what it regards as its own turf.

In 1985, however, Shmelev reports that there are 'now' some 2,000 of these bureaus and that their number is growing. Although this development is surely positive, we are told by the same source that in 1983 there were no more than 6,745 purchasing points in the cooperative network, which is little more than the number of official markets. In the future it is planned that their number should increase substantially, to reach 20,000 by 1990. A further 5,000 seasonal purchasing points are also planned, to serve the gardening associations.[84] In his speech to the 27th Party Congress, in February 1986, Tsentrosoyuz head Mikhail Trunov claimed that during 1986–90 a total of three billion rubles would be allocated for improvement of the local logistical base of his organization. That would be equal to the total amount spent for such purposes during the past 20 years.[85] As usual, however, it remains questionable to what extent these visions will actually be enacted.

From a peasant point of view, the process of marketing private output will thus most likely continue to be associated with quite substantial obstacles and difficulties. The official procurement system is not reliable. Private traders are illegal. Little help – and comfort – is available on the markets, and even the process of taking the produce to market seems like an ordeal. The quality of roads and the general shortage of transport in the Soviet economy are well-known phenomena and it is hardly surprising that most official appeals for help to be given are of little avail.[86]

More serious perhaps, is the fact that even such initiatives which are taken by the peasants themselves encounter opposition. It has long been a common practice, for example, for peasants to hitch a ride to town with passing trucks. A good sized potato on the top of a stick has been a frequent indication for drivers that it might profit them to stop and offer a ride. According to figures given in a recent *Pravda* interview, with the first deputy General Attorney of the SSSR, N. Bazhenov, and the deputy Minister of Trade of the RSFSR, V. Karnaukhov, as much as 80 per cent of the produce that is brought to the markets arrives 'on the left' (*na levo*), i.e. with drivers of government vehicles who take some extra cargo on the side.[87]

Partly as a result of the campaign against 'unearned incomes', however, such seemingly smooth solutions to the problem of scarce transport are now actively being ruled out. The consequences for the peasants are vividly illlustrated by *Pravda's* correspondent, who simply exclaims, '*Kak zhe byt krestyaninu!*' (What can a poor peasant do?). The official reaction is rather typical. Bazhenov firmly establishes that taking cargo 'on the side' is prohibited, and that society suffers great damage from such activities. Karnaukhov states that the 'speculators' must be chased off. Nevertheless, both maintain that such produce that is presently sold by the speculators must not be 'lost', and that transport 'on the side' must be carried out in a legal way. How this is to be achieved, however, is not spelled out. What we have here is thus once again an illustration of the typically Soviet attitude of wanting to get 'something for nothing'. It is fine if the private sector makes a valuable contribution, but it should preferably do so without receiving anything in return.

As a result of official unwillingness to allow private traders, the inability of the official system to compete, the lack of any positive assistance and the hostility towards private initiatives, the stage is set for precisely those consequences that have been the topic for discussion above: large prospective gains that engender corruption, and even where trade is conducted within the legal limits, profits suffic- iently high to arouse moral indignation. The deplorable outcome will be that situation of arbitrariness in the struggle against 'speculation' and 'unearned incomes', which we have characterized above as the very antithesis of the requirements of a *Rechsstaat*. Let us proceed now to look at some of the more tangible manifestations of this problem.

Local Raiding Parties

The illegality of such events that may sometimes befall Soviet peasants, on their legal way to market their legal produce, has long been a target of criticism in the press, but seemingly to little avail. In some cases, incidents may even be of such a nature that a casual observer would probably not hesitate to talk of outright highway robbery. *Izvestia*, for example, reports of a couple from Rostov *oblast* who, in a year of local drought, found themselves having to travel to nearby Voronezh *oblast* in order to buy a few buckets of potatoes for the family. On their way back they were stopped by a militia patrol and taken to a nearby procurement office where their potatoes are 'delivered' according to the *raion* procurement plan. The price paid is 30 kopeks to the kilo, or a total of 64 rubles and 80 kopeks. At the market, the couple had paid 125 rubles. To a straight question, whether the local authorities really had a legal right to act in this way, the militia sergeant answered in the negative, simply adding that he had to obey given orders.[88]

In another case, of a similar nature, a woman had visited her mother in the country and was returning to town with four sacks of potatoes. *En route* she was stopped by a militia patrol and asked for a document to prove that the potatoes were hers. This, however, she had not brought with her, thinking it unnecessary since she was not headed for the market. The four sacks were promptly confiscated, and although no law had been broken, her attempts during the coming days, to have the potatoes returned, were all frustrated.[89]

In order to cover up the blatant illegality of such actions, a number of different ploys are used. One of the more imaginative examples in this genre was reported in *Izvestia*, under the heading 'Potato Battles of the Kherson Gardeners'. Here, the *oblast* executive committee had simply declared a quarantine, prohibiting all transport of potatoes out of the *oblast*. Buses were searched and cars stopped, and anybody who was found transporting potatoes without a proper document, certifying the absence of disease, promptly had them confiscated. Curiously, however, the quarantine did not apply to potatoes transported under the auspices of the Tsentrosoyuz. These could be freely taken out of the *oblast* and sold on distant markets without risk of spreading any disease. The real purpose of the operation is clearly recognized by *Izvestia's* correspondents, who conclude that 'those special organs which are busy arranging "protective measures" least of all worry about the interests of those who are waiting to buy early vegetables on the urban markets.'[90]

In another case, which was also reported in *Izvestia*, a similar strategy was used by the executive committee of Belgorod *oblast*. A decree was issued, declaring that various potato diseases had been observed and prohibiting all transport of potatoes out of the *oblast*, without proper clearance. In this case, *Izvestia's* correspondent is even more to the point, noting that delays in the harvest of sugar beets had caused an acute shortage of transport for the procurement of potatoes. Faced with the risk of falling short of the procurement plan for the latter, a quarantine was introduced to ensure that the potatoes would at least remain within the *oblast*, where they could be procured at a later date. The seriousness of the disease was illustrated in the account of a peasant who claimed that a proper certificate for transport could be obtained, not by inspection for disease but by selling a few tsentners to the procurement office. In return, the remainder could then be taken freely to the market.[91]

A further example, in the same vein, is provided by Shmelev, who attributes difficulties in finding potatoes on the Moscow markets during the autumn of 1980 to the fact that local authorities in neighbouring areas, struggling to meet procurement plans, deployed various obstacles to prevent potatoes from being taken out of their respective areas. In one example, trucks sent out from the Moscow markets were simply stopped and the drivers forced to deliver their loads according to the local procurement plan.[92]

It is evident that this type of behaviour from local officials is intimately connected with the basic rules of the game in the Soviet system. This is clearly illustrated by the fact that it affects not only the individual peasants, but also *kolkhozy* and *sovkhozy*. Technically, the latter are allowed to market whatever surpluses they may have left, after meeting their state delivery obligations, but practice does not always follow theory. The following is Shmelev's characterization:

> Before taking produce to market, the *kolkhoz* has to 'cover' its plan. But this is not all. It not only has to wait until its own plan is fulfilled, but also until the *raion* plan is fulfilled. And should the *raion* plan be fulfilled, then it has to wait until the *oblast* as a whole has fulfilled its plan. Thus the *kolkhoz* waits, but the tomato, unfortunately, does not wait. Nor do prices on the market.[93]

This type of interference is of course different from such simple highway robbery that has been described above, but only in its outward manifestation. The act of shifting delivery obligations from failing farms onto the shoulders of those who have already met their obligations is just as illegal and has long been the target of harsh criticism in the press. Its consequences, moreover, are also similar, in that produce goes to waste and prices on the market rise. The fact that it still continues can thus be taken as a fairly firm indication of its causes being intimately linked with the basic rules of the system.

For an outside observer, it may be easy to understand that local officials may end up in situations where they are simply forced to undertake actions of the kind described above. It may even be possible to defend such behaviour as a rational necessity, given the overall rules of the game. To demand, however, that the peasants, who are subjected to these various forms of abuse and malpractice, should show such understanding is certainly a rather tall order. We shall proceed in greater detail below to take a closer look at what forms their *actual* reactions can be expected to take. Let us, however, anticipate that discussion here, by quoting a rather appropriate conclusion that was presented in an article in *Pravda*, entitled 'Why Go to the Market?'

Having surveyed a number of letters calling for stricter measures to be taken against trade on the markets, and in some cases even for them to be closed, the paper's correspondent concludes that the prevalence of such views must not be neglected:

> If such views are shared, as it were, by local officials, then it is quite realistic for a peasant, in a hired state vehicle, to be unable to make his way to the market, or at least to have no desire to attempt to do so a second time.[94]

The seriousness of the matter is further underscored, as the paper goes on to quote the fate of a group of workers from Dagestan, who had their truck stopped by a militia patrol and were forced to wait – without explanation – for a full eight days, while their load of tomatoes slowly rotted. A letter of complaint, sent to the public prosecutor, was left unanswered. Following the 1986 campaign against 'unearned incomes', the intensity of such actions by local officials has probably increased. In an article entitled 'Is the Cucumber Really Guilty?' *Pravda* surveys a number of cases where the 'struggle' is carried out in quite illegal ways,[95] and there is plenty of other evidence as well, all serving to illustrate the argument regarding the absence of a *Rechtsstaat* which has been repeatedly referred to above.

Glasnost, Perestroika and Official Attitudes

With the onset of *glasnost* and *perestroika*, those negative attitudes towards the private sector that we have outlined above have suddenly undergone a dramatic transformation – in their public presentation. We have already given some indication of how this can be traced in purely ideological pronouncements, about the 'true' nature of private agriculture, and we shall now turn to see how it has affected the official view 'from the top' regarding abuse and irregularities by local officials. At the 1986 round table discussion that was referred to above, Nikiforov led off by stating that the root of such problems lay in the fact that the private sector had become increasingly oriented towards production for the market:

> However, our economic system . . . has not been prepared for such a development, and the reaction has been one of prohibitions: to establish norms or to prohibit altogether the growing of tomatoes, cucumbers and strawberries, to reduce the size of private plots, etc. Here we must firmly underline the dangers of such a prohibitionist policy. It can bring our agriculture nothing but harm.[96]

A similarly harsh attitude was taken by Aksenii Kalinkin, a prolific writer and established authority on problems of the private sector, who led off his contribution by referring to a large number of letters received by the journal *Selskaya Nov*, from people engaged in private agriculture. The letters speak of administrative restrictions on both production and marketing, of fines levied and of growing crops being ploughed up, of plot sizes being reduced and at times even of greenhouses being torn down. Axes, saws and bulldozers are seen to be given a free play by local officials. The simple act of travelling to the market is often considered criminal, and can at times be used as an excuse for reducing or removing altogether the household's plot land. At the same time, however, neither the consumer cooperation nor *kolkhozy* or *sovkhozy* is capable of handling the produce in question. Kalinkin's conclusion is hardly a flattering one for those who are charged with supervising Soviet agriculture:

> Obstacles to marketing will produce results that are the very opposites to those intended: with the reduction in supply will follow rising prices on the markets and thus a fertile soil is created for true speculators to operate.[97]

Several other participants in the discussion presented views in a similar vein. T. Kuznetsova, for example, another well-known contributor on these matters, raised the important question of such plot incomes that exceed a 'socially normal level' (*obshchestvenno normalnyi uroven*), and firmly underlined that it has never been established what an acceptable level might be.[98] Further, a certain G. Chubukov, doctor of law at the Academy of Sciences, pointed out that while adequate legislation on private agricultural activities does exist, it is frequently displaced by administrative measures. It is firmly emphasized that the law is in no way concerned with *how* resources at the disposal of the private sector should be used, and that all restrictions imposed for such purposes, be they adopted by the *kolkhoz* meeting or by individual managers, lack a legal foundation.[99] Finally, we have the 'mandatory' *kolkhoz* chairman who is allowed to express that he finds it 'strange' to hear workers in the LPKh being called 'privateers' (*chastniki*), who fails to understand how their incomes can be 'unearned', and who is upset at the demolition of greenhouses, the destruction of crops and other such 'scandals' and 'cruelties'.[100]

These recently made statements on the role and nature of the private sector are certainly very interesting in their own right. It is rather difficult, however, to tell what should be made out of the rather contradictory picture that results if they are contrasted with that picture of grudge and harassment which has been presented in the previous chapters of this study. First of all, we may note that Nikiforov's statement about the increasing market orientation of the private sector, for which the system had allegedly not been 'prepared', is quite simply not true. As we may recall from above, the share of the private sector in the total turnover of foodstuffs fell from 30 per cent in 1950, to 10 per cent in 1965, and has since remained fairly stable. Similarly, it is of course the case that what Nikiforov refers to as a 'prohibitionist' policy is a phenomenon that has its roots in mass collectivization, rather than in a sudden and unexpected growth in private marketings during recent years. Discrimination against the private plots and the *kolkhoz* markets is endemic to the system and partly related to the expectation – or wish – that the private sector as a whole shall eventually 'die out'.

Apologetics aside, however, the question remains as to whether the current about-face in official attitudes will be sustained, or simply represents another temporary turn in the long zig-zag history of private agriculture in the Soviet Union. To give an answer to that question is of course impossible at the present time. Nevertheless, we shall return below to speculate on the considerable scepticism that we may perhaps realistically expect to find amongst the peasantry. With this we shall conclude our presentation of the 'practice of harassment', and proceed to summarize our findings thus far.

Conclusion

A recurrent theme in writings by those sympathetic to the private sector, is that of a lack of understanding of its true nature, and of an underestimation (*nedootsenka*) of its productive potential, the obvious conclusion being that increased support would result in a considerably increased contribution towards food supply. Shmelev, for example, argues that such underestimation has led to unjustified restrictions on plot activities. In order to provide some official support for this opinion, he goes on to quote the following, from the 1977 decree on support for the plots: 'However, some executive committees of the local Soviets, together with *kolkhoz* chairmen and directors of *sovkhozy* and other state and cooperative organizations underestimate these possibilities.'[101]

One explanation for the fact that such attitudes towards the private sector are widely held is certainly that many bureaucrats, who are inclined towards an easy life, consider it to be a disturbing complication, in the absence of which the socialized sector would be that much easier to manage. As Shmelev points out, such attitudes can easily turn into ingrained behaviour: 'The power of habit over thoughts and actions may explain the solidly negative attitude that is found amongst some local officials, as well as the vitality and frequency of their relapses into meddling and interference in relations to the LPKh.'[102]

This focus on bureaucratic inertia and almost instinctively hostile attitudes towards all things private will go a fair way towards explaining many of those problems of discrimination and withheld support that were discussed in the first part of this chapter. It still deals, however, with only one side of the problem. What we are left with are the *actively* hostile actions that were discussed in its latter part.

Here, as we have seen, the position of local officials is central. They are under severe pressure to meet delivery plans and will thus be sorely tempted to 'confiscate' whatever produce they can lay their hands on. If we recognize that *their* superiors in turn have *their* plan targets to meet, it may be somewhat easier to understand the obvious lack of interest in putting an end to the outright 'bandit' policies that are systematically pursued by local officials in various parts of the country.

A similar situation can be identified for the *kolkhoz* chairman, or the *sovkhoz* director, who is under pressure to meet *his* plan, and who fears that the socialized sector might fall short. He will be tempted to force the peasants to sell their produce to the farm, for further delivery according to the state procurement plan, rather than allow them to go to the market. Such practices have even been the subject of discussion in the Central Committee of the Belorussian Party, where those guilty were 'severely punished'.[103]

We shall not dwell further, however, on the possible explanations for the various forms of passive and active harassment against the private sector that have been presented above, since this would take us too far into the realm of the rules that govern the behaviour of the Soviet bureaucracy at large. Instead, the time has now come to proceed to investigate what their impact may be on the peasants as producers of food for the population.

Chapter six

Response to decline

Our presentation thus far has focused almost exclusively on the 'support' functions of the private sector. No doubt these features are the ones that most easily catch the eye, and as we have seen from our brief glimpses of the debate on this subject, they also tend to be at the centre of Soviet attention. Our account of the recent growth in emphasis on the *podkhozy* was intended to highlight the growing urgency among Soviet officialdom regarding the need for such support. Having thus established the first of those two functions of the private sector that were referred to at the outset of our study, we shall now proceed to discuss its second function, that of acting as a political stabilizer. This is of course a somewhat less obvious role, and in order to argue the case we shall have to proceed in two separate steps.

The first of these will be dealt with in the present chapter and will be aimed at investigating the impact on the peasants of those various types of discriminatory attitudes and actions that have been outlined above. Having set this objective, we must of course immediately recognize that the conceptual problems involved are of some considerable importance. If, for example, we were to focus on the Russian peasantry alone, a sense of mutual hostility and distrust between peasants and authorities is of course a phenomenon that goes way back, to times before both Lenin and Stalin. It can thus hardly be said to have resulted from the imposition of Soviet power. In this respect, important distinctions must be made between the Russian and the non-Russian elements of the population.

If, however, we start instead at the time of mass collectivization, we do find a rather natural common denominator and point of departure. In this process, all Soviet peasants were reluctantly but effectively incorporated into the Soviet economic model, in the sense of having all options other than quiet submission blocked by brute force. It is quite significant, for example, that there have since been neither rebellions nor other forms of open opposition to Soviet power from the side of the peasantry, broadly defined. It is this very fact which lies at the heart of our mechanism of political stabilization. It should be fairly obvious that forceful collectivization must have brought about very strong feelings amongst the peasantry, feelings that were directed against Soviet power. Barring open outlets, such feelings must then be assumed to have found internal outlets, intimately associated with the very system of *kolkhoz* farming. Similarly, one can reasonably assume that those various forms of discrimination and harassment that have been described above have served over time to confirm and sustain some important parts of those values and attitudes that were, so to speak, injected into the system from the very start.

In this perspective, the differences that no doubt exist between various nationalities in the Soviet population take on a somewhat reduced importance. We shall argue below that the response from the side of the peasants can be seen as a form of withdrawal, or an exit. Here it is certainly the case that typically one may expect Russian peasants to display somewhat different reactions than, say, their Georgian or Baltic comrades. The important point, however, is that although different in *origin* such reactions will be united in *function*. Most obviously, a withdrawal can of course be observed in the socialized sphere of production, in terms of slacking, drinking and general apathy. We shall also argue, however, that it has a more subtle and important dimension, in the form of a mental alienation from the official ideology and value system of the Soviet state.

While such a withdrawal may of course be loosely said to apply to broad sections of the Soviet population, the case of the peasantry is qualitatively different, in the sense of posing a more serious challenge which in the long run may even present a threat to the Soviet leadership. The second step of our presentation of the private sector as a political stabilizer will be the topic of our concluding chapter, and there we shall expand on this challenge. It will be argued that the ingredients in our mechanism of political stabilization are of such a nature that they actually present a formidable obstacle to Mikhail Gorbachev's attempt at subjecting Soviet agriculture to a process of 'restructuring', or *perestroika* as the term goes in Soviet newspeak. In this sense, the past successes in achieving political and social stability may be seen to have serious implications for the prospects of eventually reforming not only agriculture but – by implication – also the Soviet system at large.

As a general framework for the following discussion, we shall start by introducing Albert Hirschman's concepts of *Exit, Voice and Loyalty* (EVL),[1] which together make up a model for the study of different types of individual 'response' in situations of 'decline'. Its strong focus on the behaviour of individuals makes this model singularly well suited to an examination of the situation of the Soviet peasantry in the context outlined above.

The Exit–Voice Paradigm

The main ambition of EVL may be captured in the subtitle of Hirschman's seminal work on the topic: 'Responses to Decline in Firms, Organizations and States.' Briefly put, we are dealing here with situations where an individual is faced with deterioration in the quality of goods he normally purchases, or in the performance of organizations to which he belongs, and it is assumed that he can choose to respond by either objecting (Voice) or leaving (Exit). The point that is made by EVL is that the choice that is made between these two options will be of a crucial importance for the prospects of potential recuperation.

A vital assumption in this context is that of 'repairable lapse',[2] i.e. that decline in the performance of a firm or an organization need not be intentional but may rather be the result of 'organizational slack'.[3] In such a case, the decline may well be reversible, if brought to the attention of management. To an economist this assumption may be slightly surprising, as it is generally assumed that if one firm goes bankrupt (due to reasons other than falling demand) its place in the market will be taken over by another, and thus no difference will be seen. If we grant, however, that recovery is possible, then different feedback mechanisms become of interest, as signals to management that something is amiss, and it is here that Exit and Voice are seen to differ in their respective functions.

Signals to management in the form of Exit are generally assumed to be weaker than those emitted in the form of Voice, as Exit will normally occur in the form of a slow seepage. Indeed, in the special case where customers simply move between different producers in the same market, Exit and entry may cancel out, and if there are many producers, such movements may proceed for so long that no signals of decline ever succeed in reaching management.[4] In contrast, even a few outspoken individuals may create quite a lot of noise by using the Voice option, and thus perhaps also succeed in transmitting a powerful message.

The ideal situation would thus be to achieve a sufficient delay in Exit for the discontent of the individuals concerned to be channelled into Voice, which in turn sets in motion the desired process of recuperation. It is here that Loyalty – the third ingredient in EVL – enters the picture. Hirschman assumes that individuals who are possessed with a certain Loyalty to the firm or organization in question will be prepared to put up with decline longer, in the hope that by resorting to Voice they may actually succeed in promoting recuperation, and thus incidentally also justify their delay in Exit.[5] This use of the concept of Loyalty is obviously somewhat different from the everyday understanding of that word, and harsh criticism has been directed at this part of Hirschman's theory. The special nature of our application, however, allows us to avoid much of this critique and we shall refer to it only insofar as it remains relevant.

An important advantage of EVL lies in the fact that it is an unusually successful example of an interdisciplinary approach, combining economics and political science.[6] The Exit option, with its clean dichotomy, comes naturally to the economist, who is accustomed to thinking in binary terms (purchase or no purchase), whereas the Voice option, on the other hand, with its drawn out and often inconclusive perspective, comes equally naturally to the political scientist, who is used to thinking in terms of processes *per se*, rather than focus on the results of those processes. In his original book on the subject, however, Hirschman argues that the Exit–Voice paradigm is equally applicable *in toto* to both disciplines, and expresses a 'hope to demonstrate to political scientists the usefulness of economic concepts and to economists the usefulness of political concepts'.[7]

Before proceeding, it is of particular relevance for the subsequent discussion to note the context in which the concepts of Exit and Voice were originally developed. In the book where they were first presented, Hirschman describes being puzzled by the fact that the Nigerian state railways could continue a grossly inefficient operation in spite of the fact that they were exposed to competition from private trucks and buses.[8] Neoclassical economic theory would tell us that this should not be the case, since the inefficient firm would be forced to rationalize, or be driven out of the market.

Hirschman's strikingly simple conclusion was that the existence of an alternative had offered the possibility of an Exit, to those who were prepared to pay. Those who would do so, however, would presumably also be those who in the absence of such an option would have been vociferous critics of various malfunctions and shortcomings. In this perspective, Exit could be said to have *defused* Voice, thus allowing the railroads to continue their inefficient operation, without being overly bothered by protests. Many applications of this finding have since been made, in particular perhaps in the fields of education and health care,[9] where the provision of private and often highly priced alternatives leads to a substantial reduction in the level of Voice, i.e. in calls for improvements that are directed against the state run systems. A case could thus be made for suppressing competition in order to preserve the healing forces of Voice, since that feedback mechanism 'operates at its best when the customers are safely locked in.'[10]

Several objections can of course be raised against the application of this type of approach to the problem at hand – political repression and censorship inhibit the use of Voice, closed borders and a general absence of competition between the providers of goods and services preclude the use of Exit, the diffuse nature of the concept of Loyalty makes it difficult to operationalize and interpret, and the long-term nature of social processes invalidates the essentially short-term nature of Hirschman's framework. Yet, it is precisely by bringing up objections of this kind that we hope to demonstrate the usefulness of the EVL paradigm in the present case.

If we start with the problem of 'response to decline', it may actually be seen as a strength that we are *not* dealing with short-term adjustment processes. As indicated above, the concept of Loyalty is that part of Hirschman's theory which has drawn perhaps the harshest criticism from reviewers. A. H. Birch, for example, has argued that Hirschman has the correlation between Voice and Loyalty going the wrong way, since Loyalty ought to mean a 'disposition to accept rather than a disposition to criticize', whereas Michael Laver sees the concept of Loyalty as falling between two stools, being 'probably the most self-contradictory part' of the whole theory.[11] One of the perhaps most serious criticisms, however, that has been directed against the EVL theory has been voiced by Brian Barry, who has argued that Loyalty in Hirschman's use is no more than an '*ad hoc* equation filler', with no independent theoretical role to play.[12]

Barry's main objection is that Loyalty does not capture a 'real social phenomenon', and it is here that we shall argue that our application largely escapes those criticisms that have been directed at EVL in its original formulation. While the perspective used by Hirschman is essentially one of short-term utility maximizing adjustments, ours is that of a more long-term process of socialization of individuals within the sphere that is controlled by government policy. Their 'response' should thus be seen not as short-term adjustments in behaviour, but rather as a more long-term formation of attitudes and values, and a corresponding conditioning and adaptation of actions.

What we are dealing with here is actually fairly close to that 'real social phenomenon' which was originally found lacking by Barry. The long-term nature of the process at hand also warrants making the observation that even if the external forces of 'decline' are reversed – which has clearly been the case with respect to official policy towards the private sector during the last decade – it may be quite some time before the nature of the previously determined response is accordingly adjusted. The dynamic aspects of the process are thus of crucial importance and shall be at the centre of our attention below.

The real core of the presentation of our mechanism of 'political stabilization' will be to demonstrate that the concept of Loyalty, as understood by Hirschman, does indeed reflect a social reality with important implications for peasant attitudes and actions. In the present chapter we shall pursue that argument at the micro level, by focusing on the educational role that we have seen ascribed to the private sector by academician Nikonov. In our concluding chapter, we shall then proceed to the macro level, to look at the role and interaction of official ideology and popular expectations.

Before embarking on that discussion, however, we shall have to examine in greater detail our own understanding of the concept of Loyalty. First of all, it is certainly the case that Loyalty – in ours as well as in the original Hirschman sense – is a phenomenon that can be neither observed nor measured. In essence, it is nothing but a convenient shorthand for a certain type of social and economic behaviour. Perhaps sociology might have a more suitable terminology to offer. There are two good reasons, however, for sticking with the original formulation. One is the simple fact of the broad currency that it has gained, thanks to previous works by Hirschman and others, while the second concerns its clearly defined instrumental role in that triad which also includes the more straightforward concepts of Exit and Voice. Given some reservations, which will be spelled out in greater detail as we go along, our following presentation shall thus assign an important role to Loyalty, 'Hirschman style'.

A further point of importance, in this context, is that Loyalty as a social phenomenon, in both ours and in its more commonly accepted sense, can be described in terms of what Jon Elster has referred to as 'states that are essentially by-products'.[13] Classic examples used by Elster are faith, courage and humility, all of which represent states of mind that cannot be 'willed'. By direct action one may certainly succeed in producing religion, bravado and meekness, but these are of course nothing but substitutes for the real thing. The difficulty, however, is that in their outward appearance the two sets may be deceivingly similar, and that in some cases one may successfully engage in what Elster refers to as 'faking'.

The reason for making these reflections is of course that the problems at hand are intimately linked with the problems of understanding and interpreting Soviet ideology in general and the creation of 'Soviet Man' in particular. By means of the very elaborate Soviet machinery of agitation and propaganda – the *agitprop* – and with a system of political penalties and rewards which promotes conformity, it will certainly be possible to instill in Soviet citizens an outward lip-service to the desired set of values and ideals. It may even be plausibly argued that as far as the rank-and-file are concerned, the

Party's ambitions stretch no further than that. Faking will in this context be sufficient.

The really crucial problem at hand, however, concerns the question of what states of mind will be produced as a *by-product* of direct action. It is in this context that our discussion of Loyalty should be placed. Let us start by recognizing that we are dealing with two essentially different kinds of action. In addition to normative *agitprop* statements about the nature of the utopia, there are also certain practical policy measures which will affect individuals in some unspecified way. The interplay between these two is the real driving force in a dynamic sense.

If there is reasonable coherence between *agitprop* and perceived reality, a positive spiral of cumulative causation may then be set in motion. As a by-product, economic and social policy measures will produce such sentiments that cannot be directly willed, and if these correspond with the message of *agitprop*, then the relevance and credibility of the latter will increase. Consequently it will be able to fulfill the purpose of *agitprop* in the original 'best' sense of that concept, i.e. to mobilize and spur the individuals on towards better deeds for the common good. If, on the other hand, there is a large gap between the utopia and perceived reality, a reversal of that process is a distinct possibility. Practical policy measures will produce sentiments that are seriously out of tune with *agitprop*, and as the latter loses in relevance and credibility its message will in turn serve to deepen the sense of alienation. Thus productive effort by the individuals will be further stifled. A negative spiral has been established.

The purpose of our previous presentation of negative ideological pronouncements and outrightly hostile actions against the private sector has been to provide a picture of such actions – and inactions – by the Soviet state which have served over time to produce certain 'states' in the minds of Soviet peasants. These states, which will now combine to make up that composite which has been broadly termed 'Loyalty', will acquire an important instrumental role in the peasantry's 'response to decline'. We shall start our account by looking at the component parts of that response, i.e. such actions or inactions by the peasants that are conditioned by the various states of mind.

Soft Exits

In order perhaps better to understand the importance of what we shall refer to below as 'soft Exits', we shall make use of what Stephen White has called 'mechanisms of adaptation'. In a situation where deteriorating economic performance is starting to pose a threat to the legitimacy of a communist regime, argues White, the authorities can in various ways 'absorb and process' demands from the population and thus 'perhaps preempt demands for more far-reaching and antisystemic change.'[14] What is proposed here is strikingly reminiscent of our respective mechanisms of Exit and Voice. By seeking in various ways to incorporate the individual into the system, a certain type of Loyalty (Hirschman style) will be promoted which serves to delay Exit and thus hopefully also to activate constructive forms of Voice – in terms of participation. Examples given by White are *electoral linkage*, i.e. a higher degree of an at least apparent choice in the political process, *political incorporation*, which refers chiefly to a broadening of membership in the ruling Party, *associational incorporation*, which serves to channel political energies via official bodies and organizations, and finally the writing of *letters*, to the media as well as to Party and state organs.[15]

The essence of what has been said in our previous chapters, however, presents the very opposite of such incorporation. We have seen how not only proper peasants but increasingly also members of the urban population are granted small islands of their own, where private activities for private gain can be pursued, inside the socialist and centrally controlled sphere. Although the Soviets may prefer to express themselves in normative ideological statements, they are also, as we have seen above, clearly conscious of the dangers that are involved. In a social and psychological sense, we shall interpret such activities as a mental Exit from the state-controlled sphere of the system, with all the adherent implications derived from a generally repressive setting. This Exit forms an important part of our mechanism of 'political stabilization', as it provides a safety valve for the release of frustration and discontent with various aspects of daily life and work inside the state-controlled sphere.

At the same time, however, it does not represent a 'proper' Exit, in the original interpretation of that concept. Instead we shall speak of a 'soft' Exit.[16] This will raise two important distinctions. First, if we consider plot and market activities, it is clearly the case that superficially no Exit at all is involved. The people concerned remain in place and the fruits of their labour will in the final analysis benefit the system as a whole. What does take place is a more subtle process, which we can characterize as an Exit from one room (the socialized sphere) and an Entry into another (the private sphere), both of which are located within the same building.[17] This we shall define as a soft *productive* Exit. It is an Exit in the sense that mentally people leave the socialized and controlled sphere of Soviet society. It is soft, however, in the sense that physically they continue to work for it, albeit indirectly.

Our second case is a considerably more serious one. It concerns activities such as drinking, slacking and general apathy and is different in the sense that the mental Exit from the system is not accompanied by any productive effort. On the contrary, we may plausibly argue that it is actually connected with a certain destruction of human capital, in terms of alcoholism, nihilism and a general erosion of discipline, responsibility and moral obligations. This we shall refer to as a soft *improductive* Exit.

While different in form, these two types of soft Exit are united in function, in that they serve as a safety valve for the release of frustration and discontent. Moreover, much of the resource use in the private sector would not be reallocated to the socialized sector if a more restrictive policy were introduced towards the plots. Hence it may be concluded – as a first approximation – that the regime gains in two ways from allowing an Exit into private sector activities. It gains in *economic* terms, by mobilizing otherwise idle land and labour resources. It gains in *political* terms, by defusing some of that political instability which might otherwise have been triggered off by a discontent with the present performance of the official system of production and distribution of foodstuffs. Using the terminology proposed by academician Nikonov, the former gain covers not only the *economic* function of increased food supply, but also the *social* function of mobilizing labour and of generating extra income. What has been referred to above as political gains is of course not officially recognized.

This first approximation should, however, be contrasted to the long-run consequences that have been repeatedly hinted at in the preceding chapters. The really crucial question to be raised concerns Elster's discussion of 'states that are by-products', i.e. what impact the policy of grudging support in combination with periodic harassment may have on the way in which the individuals concerned perceive their present as well as their future situation within the Soviet system. As we have seen above, we are dealing here not only with the bulk of the peasantry, but also with a rapidly growing segment of the urban population. The question ought thus to be of some considerable importance.

Our following argument will proceed in three steps. First, we shall focus on that process of consistent discrimination which has served over the decades to 'depeasantize'[18] Soviet agriculture, by severing the attachment of the tiller to the soil. Then we shall investigate what nature of reactions may be provoked from the side of the peasants to this discrimination, i.e. what type of 'response' has resulted, and finally we shall study the entry of the urban population, which presents by far the most peculiar ingredient in the picture. Let us start with the legacy from the past.

Discrimination

There are certainly ample grounds for a sense of discrimination to be experienced by the Soviet peasantry. There are quite a few obvious signposts, as reminders of repeatedly frustrated illusions of a finally satisfied land hunger. The 1861 emancipation of serfs gave the Russian peasants 'their' land but also brought heavy redemption payments which in essence meant a continuation of serfdom in a new guise. The uprising of 1905 was a product of peasant land hunger, but again hopes were frustrated. The revolutions of 1917 gave the peasants 'their' land a second time, by way of the spontaneous land reform that followed the decree on nationalization of all land, but it also brought War Communism and *prodrazverstka*, the policy of forceful requisition. NEP provided a new promise for the peasants of finally getting 'their' land, but this ended in the horrors of mass collectivization, in 1930–32. The upswing of the private sector in the years after Stalin's death, finally, gave new life to the illusion, but as we know that ended in Khrushchev's attacks, in 1958–60 and in 1963.

Although much of this concerns Russian peasants and older generations, if we think in terms of a 'collective memory' and a common subsequent fate, it should be rather obvious that these various experiences must weigh heavily on the mentality of the current generation of Soviet peasants, and perhaps not only on them. As we have seen above, a very large part of the Soviet population is still classified as 'rural', which means that an overwhelming majority of the urban population must be either first or second generation emigrants from the rural sector. In 1982, *Komsomolskaya Pravda* wrote that 'more than half of the current urban population are yesterday's countryfolks, united by family, economic or other ties with the places where they were born.'[19]

Moreover, in a highly interesting article about peasant attitudes to the soil, which was published towards the very end of the decaying Brezhnev leadership, Soviet writer Ivan Vasilev, who is another valiant champion of the private sector, offered the following description of this newly urbanized section of the Soviet peasantry:

> The peasant children have been educated, they have become engineers, teachers, doctors, artists, journalists – they have become intellectuals, but bearing with them psychologically their peasant origins. Psychology does not change by far as fast as does an entry in the column of 'social status'. Although gradually weaker, it lives on in children and grandchildren.[20]

This peasant heritage of the urban population is of obvious relevance for our discussion of the private fringe, and we shall have reason in a moment to return to it. First, however, let us look at the attitudes and self-respect (or lack thereof) that can be found among the proper peasants. Since we are obviously dealing here with a spectrum of emotions, we shall pin down two extremes, representing very different situations and implying equally different problems for the political leadership to handle.

Writing about the effects of the introduction of the NEP, at the beginning of the 1920s, Moshe Lewin maintains that 'its policies probably gave many peasants a sense of social promotion as millions of them became *khozyaeva*, i.e. independent, respected and self-respecting producers in their own communities.'[21] This can be compared with the effects of Stalin's mass collectivization, in relation to which Conquest quotes the following from a novel that was published in Moscow in 1934: 'Not one of them was guilty of anything; but they belonged to a class that was guilty of everything.'[22]

Seen from a peasant perspective, these two examples no doubt represent quite extreme opposites, the former promising milk and honey and the latter having nothing but a 'second serfdom' to offer. Seen from the perspective of Party ideology and official attitudes, however, we have a radically different situation. If we think in terms of a desire to control the peasants for purely political reasons, as was indicated by Yanov in his above statement about the *kolkhoz* system, then the two situations differ only in terms of the amount of violence that was used in order to reach the target.

In this perspective, the former represents 'too little', as NEP allowed an open Exit by the peasants from the state-controlled sphere, with Party control over the vital exchange between town and country being reduced from direct administrative orders to a diffuse, and perhaps also poorly understood, market mechanism. The latter, on the other hand, represents 'too much', as collectivization produced a soft improductive Exit into drinking, slacking, apathy and attempts at escape, all of which were obviously costly in terms of output forgone. Both extremes can thus be seen to have their own drawbacks, but they also share one important common feature, in terms of an official distrust of the peasantry.

The ideological continuity that exists on this point may be illustrated by the very simple facts that it would not be until 1981 that the peasants were generally issued with domestic passports, without which they could not leave the farms even temporarily,[23] and that it would not be until the 1988 *kolkhoz* charter that they were given formal rights to actually opt out of the *kolkhoz*, i.e. to give up their membership.[24] The consistent experience of always being last in line for resources would also serve to drive home the message. No doubt, it is such features that have prompted Yanov to talk about agriculture as being a 'Party affair', incapable of being left to its own devices.[25]

Granted this desire for control, and viewed against the background of the respective 'costs' involved in the two extremes outlined above, there would, intuitively, seem to exist a preferred middle way, that satisfies the desire for control at a minimum cost in terms of initiative and output forgone. That this actually is the case will be argued in the final section of this chapter, where we shall present a composite picture of the various forms of peasant responses. This picture shall be derived from our concept of passive Loyalty and it will be seen to fit precisely the demands of a 'middle way'. Before doing so, however, we have quite a few points left to discuss on the subject of soft Exits.

At this point, it may be relevant to recall the largely rhetorical question that was raised by Shmelev above, as to whether we are dealing here with 'millions of private entrepreneurs, having treasured deep in their souls individualist interests and leanings.' Since the answer to this question is of central importance to our argument, we shall have to proceed here in somewhat greater detail.

First of all, it should go without saying that many of the ideologically tinged arguments about 'private property instincts' and 'petty-bourgeois' leanings must obviously be very far from empty declamatory statements. It must surely be the case that the plots have had precisely these types of effects, on at least some parts of the peasantry, given regional variations in the conditions for profitable plot activities. The simple fact, for example, that the protagonists of the private sector go to such lengths in their ambitions to refute the existence of these kinds of attitudes, without discussing the really central issues involved, forms a strong ground for suspecting that there is indeed some fire behind the acrid smoke.

If we return to White's perspective of 'mechanisms of adaptation', i.e. the incorporation of the citizens into the official sphere, and their socialization into its norms and value systems, it would seem reasonable to wonder to what extent the peasants really feel that they are part of Soviet society and its official ideology. The following section will argue that a peculiar process of 'disconnection' of the peasantry from Soviet society at large has taken place, with serious consequences for the performance of both the socialized and the private agricultural sectors, a disconnection which is reflected *inter alia* in Yanov's presentation of agriculture as a 'Party affair'.

A Disconnected Class

In one of his books on the subject, Shmelev takes issue with an anonymous opponent who had allegedly talked about the Soviet peasant in Faustian terms, as having two souls – one a participant in production in the socialized sector, and the other a private (*chastnyi*) smallholder: 'Ah, there lives in my breast two souls. One a stranger to the other, and both striving to be separated.'[26] This view is ridiculed along the lines of such rhetorical questions and normative ideological statements that have been repeatedly quoted above, but the arguments are largely declamatory and not particularly convincing.

A similar, and considerably more sophisticated, view is presented by Vasilev, who sees the peasant nature as twofold – based on 'work' (*trudovoe*) and on 'private property' (*chastnosobstvennicheskoe*), or 'proletarian' (*proletarskoe*) and 'petty-bourgeois' (*melkoburzhuaznoe*). His argument is that the process of erasing class distinctions in Soviet society works in the direction of suppressing the latter – negative – side, and the way in which this is done is likened to the way in which the peasant fights weeds on his fields, i.e. by creating conditions under which they cannot survive. The field where the 'psychological weeds' grow, in this case, is the economy and it is thus only via changes in the economic sphere that the psychology of the individual can be influenced.[27]

A striking illustration of such a change, which is highly relevant in our context, is the important 1966 decree which introduced an element of guaranteed pay in the *kolkhozy*.[28] Here Vasilev argues that while this measure was economically and socially necessary, in order to increase production and erase class distinctions between worker and peasant, it left one vital aspect without consideration, namely 'how to preserve, under the new form of remuneration, the previous dependence of the individual on the final results of his work?'[29]

The conclusion that is drawn by Vasilev forms a devastating condemnation of an agricultural policy which has allowed the spread of a 'weed' of the most dangerous kind – a peasant who is indifferent to the soil that is supposed to feed him:

> The soil as a source of material wealth could no longer hold him, and he who did stay started looking at it like a peculiar form of factory workshop, in which to 'chase norms'. Briefly put, no true involvement in the needs of production was promoted, rather the opposite. And the soil answered, as it would have to answer, via falling yields. Thus arose a difficult food problem.[30]

However, this type of disconnection of the peasants from work in the socialized sector is far from the whole story. In a narrowly economic perspective, we can certainly present a picture where increases in socialized sector pay, in combination with strong income effects, derived from the strenuous nature of work on the plot, lead to a reduced interest in plot activity, while the absence of a nexus between effort and rewards in the socialized sector produces the indifference towards the soil that is outlined by Vasilev.

In order to create a fuller understanding, however, we ought also to take into consideration the psychological effects on the peasants of what has been discussed above. The persistent feeling of being at the bottom of society, the notorious difficulties in obtaining inputs, and the periodic harassment on the way to market, must certainly have the effect of making many peasants give up not only on making an effort in the socialized sector, but also on putting effort into work on their own plots. The latter would apply in particular to animal husbandry, the traditional pivot of the private sector. As we may recall, this is becoming an increasingly unattractive activity for both the youth and those with education, while families with a *babushka* form its mainstay. Such a situation would hardly seem to warrant predictions for future expansion, or even recovery.

An extreme illustration of what may result from this 'double' disconnection of the peasants, from socialized as well as private soil, is provided by an article in *Literaturnaya Gazeta*, from 1982, which tells the story of a village that is more or less completely given over to various handicrafts. The items made, however, are poor 'caricatures' of works by the 'old masters' of Russian handicrafts, sold at 'speculative' prices, to tourists looking for keepsakes. The author of the article in question is noticeably upset with these 'bunglers', who are 'speculating on the good reputation of the true masters', but from our perspective it is not so much the quality of the art that is important, as the impact on peasant interest in agricultural activities:

> Considerable harm is done. One *kolkhoz* chairman completely shamelessly told me that about half of the *kolkhozniki* took no part whatsoever in agricultural work, during neither summer nor winter, preferring instead to sit at home making souvenirs.[31]

The picture that is presented here is one of a bustling activity in all fields except agriculture. Many *kolkhozy* have stopped thinking about animal husbandry, fertilizer and gardening, focusing instead on the manufacture of pseudo-souvenirs. Lathes are found in thousands of homes – young and old are grinding, polishing and painting. Most importantly perhaps, a great wealth is amassed, a wealth which makes earnings possibilities in agricultural work – socialized as well as private – fade into insignificance. The account is obviously not representative, and it is certainly the case that the author feels he has an axe to grind. Nonetheless, it is hardly a fiction of imagination and can thus be taken to reflect a situation which might become more prevalent if the current rot in the socialized sector of agriculture cannot be overcome.

In commenting on this and on other similar articles, Vasilev focuses on the common psychological denominator that lies in the moral consequences that must result from the spread of such 'mercenary attitudes' (*torgashestvo*). The roots of the evil are seen to spread in a number of different directions – in the unrealistic nature of state plans, in the monopoly position of agriculture's suppliers, in the distortion of prices between agricultural produce and souvenirs, and between bread and flowers, in the erosion of personal responsibility, etc., but there is one factor which is seen to stand out:

> To my mind, the core of such reasons lies in the twofold nature of psychology. Pressing economic necessities (let us express it thus), like plans, funds and prices, all work like a ray of sunshine on the suppressed weed, on that side of psychology which is not connected with work, allowing it to assume threatening proportions, and causing entire collectives to be infected with private property instincts and a mercenary spirit.[32]

Vasilev returns here to that same argument on the twofold nature of peasant psychology which was referred to above. While previously, however, he was seen to hold that official policy served to suppress its 'petty-bourgeois' side, by creating conditions under which the 'weed' of individualism could not grow, here we can see that he is also clearly aware of the fact that in many cases the actual conditions facing the peasants may well be precisely those that are conducive to growth of the 'wrong' side of their psychology, i.e. that side which is connected with individualist ideals and inclinations. Against this background, one would certainly not hesitate to agree with his repeated statement, that what is needed is a 'return to the soil' (*vozvrashchenie k zemle*), but it is clearly debatable to what extent current official policy actually promotes the attainment of this objective. From the presentation in this and earlier chapters, it should be obvious that our understanding is that of an opposite influence.

Serious as it may be, however, this process of depeasantizing Soviet agriculture is unfortunately still not sufficient in order to present the full picture. If we return yet again to White's perspective of adaptation and incorporation, we shall see that there is a further aspect of the problem, in addition to indifference to the soil, an aspect which derives from the Exit into private sector activities and which serves to compound the consequences of the disconnection from the official sphere of society.

Many of those Soviet sources that are sympathetic to the private sector emphasize the long hours, and the heavy work, that are involved in operating a private plot – in addition to working full time for the socialized sector. Shmelev, for example, has the following to say:

> We cannot agree with those who categorically judge incomes from trade on the market as 'unearned'. Behind such trade, as we know, lies work on the household plot (hard work) and also difficulties in the actual process of marketing.[33]

A similar position is taken by Dyachkov and Sorokin, who point out that activities in the private sector are based chiefly on hard, manual labour, and are consequently of low productivity. According to a source cited by them, labour time expended per hectare in the private sector is 8 times higher than in *kolkhozy* and 18 times higher than in *sovkhozy*.[34] Given the very different nature of chores performed in the two sectors, this comparison is obviously misleading if seen in narrow economic terms. Seen in a social perspective, however, it does underscore the moral rights to rewards for hard work, and in this light it is of course of some considerable importance. Statements of this kind are no doubt made in order to refute accusations against the private sector of being an independent 'private' (*chastnyi*) sector, with all the negative connotations of 'speculation' that are carried by that word.

Of particular interest in our context, moreover, is the fact that labour intensive plot activities will not only have a *direct* influence on peasant mentality, but will also have an *indirect* influence, in the sense of crowding out alternative uses of time. The latter has been captured by Lewin in the following way:

> The serious activity on the plot thus added a burden of working hours which other classes did not share and deprived the peasant of many opportunities to broaden his cultural horizons. Modern Soviet sociologists have grasped the full implications of this factor. Some have pointed out that the private plots of even the *sovkhoz* and industrial workers, still widespread and economically quite important, influence the outlook and mentality of those involved.[35]

A precondition for any successful form of incorporation of the individuals into the official sphere of the economy, and thus into its official ideology and value system, is of course that these individuals must have time to absorb the contents of the message. As we may recall from our previous discussion on Soviet attitudes to the private sector, one of the voices quoted there indicated the importance of precisely this problem, by pointing out that the future of Soviet agriculture must rest on the shoulders of young people, who are freed of work on the plots in order to have time for books and television. The actual situation, however, as was indicated by Lewin above, is unfortunately very different. Consequently, the prospects of actually 'reaching' the peasants are rather limited, even if they should be interested in being reached in the first place, which of course is debatable.

In this context, it may be relevant to recall from above the stress that was placed by academician Nikonov on the 'educational' function of the private sector. This point is repeatedly stressed by a number of other Soviet sources as well. Shmelev, for example, argues that 'we can and must view the LPKh as a condition for education and strengthening of working habits and positive attitudes to work, as a direct opposite to idleness.' With regard to children in particular he sees this function as important:

> Personal auxiliary agriculture helps familiarize children of peasant families with work. There they acquire their first working experiences, in caring for the animals and tending to the crops. In rural families having no plot, we often observe the emergence of a cult of consumerism, and no love for the peasant way of life is instilled from young age.[36]

A really crucial implication of this focus on the educational function of the plots concerns the central issue of *why* one should have a plot in the first place. If the motive is chiefly money, then all those obstacles and difficulties that have been outlined above will be relevant only in determining the level of market prices that the peasants regard as constituting adequate compensation. In this case we would expect to observe a prevalence of such forms of specialization in profitable crops that were presented above as being ideologically suspect. The issue of 'education' will then be of greater relevance to the various forms of 'mercenary' activities, than to a 'love for the peasant way of life'. It will serve to spread precisely that infection of 'private property instincts' which was indicated by Vasilev above, and it is of course particularly serious if this is done at an early age.

If, on the other hand, the motive for having a plot should derive chiefly from a preference for the 'peasant way of life', and thus be focused more on supplying food for the family than on production for the market, then we have the reverse side of the coin. In this case, the various forms of harassment and obstacles will be of considerable importance in shaping peasant mentality in a more narrow understanding of that concept. Given the difficulties of purchasing food in rural areas, the decision to produce food for the family does not really involve much of a choice. Most of those young people who grow up on the farms will consequently take an active part in work on the plot, and will thus continually be subjected to such discrimination that has been described above. In the process, they will be 'educated' into the 'traditional' role of Soviet (and Russian) peasants – that of being relegated to the very bottom rungs of society.

This is something very different from what Shmelev promises, in calling for an increased support for the private sector: 'Such a policy on the LPKh will strengthen the peasant's conviction, that his work on the plot is of great social and economic importance to society.'[37]

A compelling conclusion to be drawn regarding the 'educational' role of the private plots is thus that irrespective of the underlying motive we may suspect that activities in the private sector will serve to reinforce peasant alienation from the ideals and values of official Soviet society, rather than promote that 'return to the soil' which is sought by Vasilev. If the current 're-evaluation' of the role of the private sector should actually prove to be the start of a basically new policy, and if it should prove possible to force local officials into adapting to this new policy – both of which are very big ifs – it will still take a rather long time before peasant attitudes and actions are accordingly adjusted. Meanwhile the 'new' policy will certainly come under heavy flak.

In passing, we should perhaps note that a somewhat different situation may arguably arise in the case of the private 'fringe'. Here urban dwellers who are taking up work on a plot are allowed not only to escape from a difficult food situation in the cities, but in the process also to enjoy healthy and meaningful recreation. We shall have more to say about this argument after looking briefly at the overall picture of such forms of 'support' that are rendered by the urban population at large.

Replacements

The consequences of that process of 'disconnecting' the peasants from the soil which has been described above have been given ample illustration in the track record of Soviet agriculture during the past decades. For a Western farmer it may perhaps be a bit hard to understand that even during times of peak labour demand Soviet 'peasants' normally start work at 8 a.m., that the farmer has already been in the fields a couple of hours or more when the *kolkhoznik* rises for breakfast, and that he may still be out there, profiting from the last hour of summer light, when the *kolkhoznik* is already lounging comfortably in front of his television. In order, however, to understand what has been stated above it is no doubt explanations of precisely this kind that are called for.

With the proper peasants being thus disconnected, others will have to step in, and first in line here are those 15 or maybe even 20 million people from the urban population who are sent into the fields every summer to 'help out', as we have indicated above. Such rescue operations are normally known as *shefstvo*, an expression which derives from the French word *chef* and which thus has clear connotations of someone else assuming command. The following interpretation of its actual meaning is offered by Yu. Chernichenko:

> By *shefstvo* we understand an extra-economic transfer of labour and material resources from the non-agricultural branches of the economy to agriculture. It is an activity which is unplanned (at least it is not included in the five-year plan) and which is very poorly controlled by the financial and judicial authorities.[38]

As a general framework for these operations, central decrees are issued which give official permission to 'enlist' both vehicles and urban dwellers for work on the farms at times of peak demand.[39] Pay for work thus performed is made according to norms and rates applicable in each case, and in addition 75 per cent of normal average earnings is to be paid by the regular employer of the draftees. The latter payments have to be made out of the current enterprise wage fund, and we can therefore expect industrial management not to be overly enthusiastic about rendering this form of assistance.

It is difficult to arrive at the exact extent and nature of these rescue operations. Since they are not provided for in the plans, no official records are kept. We thus have to rely on isolated observations or on a painstaking study of scattered evidence published in various central and local papers.[40] A general impression is that rural villages and small towns close to the farms are swept with a fine tooth comb for virtually all available able-bodied workers, while industries and institutions in the larger cities are subjected to more reasonable 'taxation'. Not only industrial workers are involved, however, but people from all walks of life, even including foreign students. One reason why such help is needed is of course the poor level of mechanization of field operations. A more serious factor, however, ought to be what we have referred to above as a disconnection of the tiller from the soil. The latter would seem to provide at least part of an explanation for not only the poor quality of work, but also the poor care and utilization of existing machinery. It is also indicative of the rapid spread of this moral rot that the latter problem seems to be a growing one.

The quality of the 'help' that is thus rendered is at best anecdotal, and it is obviously far from sufficient to make up for the disconnection of the proper peasants.[41] It is here that the urban population enters the picture, on its own little fringe. That the official attitude towards such activities has recently turned favourable is well in line with our presentation of agriculture as a Black Hole, always ready to absorb new resources. As we may recall from our previous account, the growth of the fringe is at present very rapid. An important explanation for the interest in plots and dachas is surely to be found in that deterioration in urban food supply which took place primarily during the 1970s, and the rapid growth as such should probably be seen more as a result of a sudden release of a built-up demand than as an increase in that demand itself. As long as the official policy continues to be favourable, it thus seems likely that the fringe will continue to grow, at least throughout the remainder of the present decade and perhaps well into the next. Apart from food supply, moreover, there is also another interesting aspect which might be added here, in the form of a rapidly increasing inflationary overhang.

In an article on the consequences of the latter problem, *Pravda* concludes that what is needed is 'new spheres for spending money.'[42] Several such spheres are listed, including an increased output of cars and the promotion of cooperative housing construction. Of particular interest for our purposes, however, is the emphasis that is placed on gardening associations and similar arrangements. It is thus not only the case that there is a major problem of food supply in the cities. People living there also have surplus cash to spend. Both of these will of course work in combination, adding to the already large pressure of demand for garden plots.

By catering for such demands, the authorities may certainly succeed in somewhat raising the real value of the Soviet currency, and thereby marginally restoring some of the incentive effect of money wages. The current emphasis on *dachniki* and gardening associations should thus be seen not only against the – admittedly important – background of welcoming additional agricultural output and increasing the welfare of the urban workers, but perhaps also as a way of absorbing some part of the rapidly growing holdings of surplus cash. All of these will combine to produce a favourable official attitude towards satisfying the demand for plots.

It must be realized, however, that this process of attracting the urban population into agriculture, whilst quietly allowing the rural population to abandon it via various forms of soft Exit, may result in a rather peculiar outcome of reversed responsibilities. As the urban population assumes a greater responsibility for feeding itself, the support offered to 'proper' agriculture is – by implication – that of permitting an increased scope for soft Exits. As the latter increase in importance, moreover, the need for support will also increase. The two processes can thus actually be seen to reinforce each other, in a rather unfortunate way. The most important point to note in this development is its impact on the people involved, above all the youth. In this respect, Vasilev touches on an important issue when he claims that the private sector is presently assuming an entirely new function, in stimulating a 'psychological return of man to the soil'.[43] The background to this claim is that process of reversal which was described above. With the move of the rural young into the cities and with the drafting of reluctant urban workers into agriculture, the stage is set for precisely that unfortunate indifference to the soil which forms the *leitmotif* of Vasilev's presentation and which activities on the plots are now expected to help cure.

The issue at hand is obviously a rather complex one. If we think in the narrow terms of what is currently presented as official policy, then it is certainly correct to say – as did Vasilev already back in 1982 – that there is under way a process of 'reevaluation of the role of the personal sector'. At that time, Vasilev's understanding of the 'reevaluation' must have been rather controversial, since he considered that plot activities were no longer associated with the old familiar problem of 'private property instincts'. Instead, they were credited with stimulating diligence and conscientiousness, the point being that if you learn to be a good 'master' on your own land, then you will also make a better worker for the common good. Today, as we know, such views are no longer on the cutting edge of the reform debate.

There are several important dimensions involved here, not the least of which concerns the educational function of the private sector and the question of *why* this change in attitude should suddenly have occurred. Let us start, however, by thinking in the somewhat broader terms of by whom and for what purpose the alleged return to the soil is taking place. This calls for making a distinction between the proper plots and the private fringe. The evidence presented above would indicate that if we think purely in terms of the number of people involved, then the real 'return to the soil' is presently being made by the urban population on the private fringe rather than by the proper peasants. Vasilev's emphasis on a *psychological* return, however, indicates that this is not what he has immediately in mind. Let us therefore start by looking at the situation in the core of the private sector, i.e. on the proper plots.

While it is no doubt true that the private sector as a whole is presently receiving greater attention in official decrees and pronouncements, there is considerable uncertainty concerning (a) the extent to which support is actually materializing, and (b) the potential response to this support if it were forthcoming. In the case of the proper plots, we have decades of historical experience suggesting that a relaxation of restrictions and some minor encouragement can produce both quick and important results. It is undoubtedly the case that at the heart of the current 're-evaluation' of policy there must lie an expectation that a relaxation of the ideological stance against private activities can be traded in for some form of a bail-out to be performed by the private sector. It is doubtful, however, whether this historical experience is directly applicable to the present situation.

It is a fact, for example, that the period 1977–81 showed very little – if any – improvement in performance that could be ascribed to the kind of support that had been promised in the 1977 decree.[44] *Inter alia*, this is reflected in the appearance of the 1981 decree, which repeated much of the same message. Although the first half of the 1980s has presented a somewhat better picture than 1976–80, it remains an open question as to whether this has been due to temporary weather factors, affecting the critical supply of feed for private livestock, or whether we are observing a sustained trend. As Wädekin has shown, the picture varies greatly between different regions, and, particularly with respect to livestock numbers, demographic factors may be considerably more important determinants of performance than changes in the official attitude.[45] Sociological evidence on the attractiveness of animal husbandry that has been presented and discussed above would also seem to support such a view. It thus appears legitimate to wonder whether the private sector really holds any remaining hidden reserves that can be mobilized on a scale that corresponds to the current level of official attention.

In addition, there is also uncertainty concerning the extent to which local officials would actually allow an expanded role of private production and market trade. As we know, they have quite a latitude for obstruction in such matters. On the one hand, the need to impose sanctions against such interference has been recently underscored,[46] but on the other we have also seen that the 1988 *kolkhoz* charter actually delegates decisions on the extent of private activities to the local level. It thus remains to be seen whether any significant change will result on this count. All in all, if the expected 'payment' for a more benign official policy should fail to materialize, the current 'reevaluation' of the role of the private sector might actually turn out to be counterproductive. Those who are against private activities, and who are currently absent from the pages of the press, will find justification for their beliefs. The consequence would certainly be a sharpened ideological conflict.

This leaves us with the case of the private fringe, where the situation is somewhat different. To the extent that we are dealing here with new 'recruits', there may well be an inclination in the short run to take delight in the newly acquired possibilities for recreation and vegetable growing rather than to brood over the problems of harassment and poor support. Moreover, since their production will be largely oriented towards own consumption they will not suffer from the full range of the problems that have been outlined above. Accordingly we can expect here, initially, to find the full benefits of the soft *productive* Exit.

Over time, however, these people will also in various ways be incorporated into the complex of problems facing plot agriculture. The evidence presented above on the situation of the gardening associations and the *dachniki* would indicate what might lie ahead in this respect. The point about an extensive peasant heritage of people active on the private fringe would also seem to be relevant in this context. For these people, impressions of the 'collective memory' of the past lot of Soviet peasants will be blended with personal experiences of the current situation of those active on the fringe. Hence, if we think in terms of a group of initially positive – or at least neutral – urban dwellers who are being offered private plots, over time the 'educational' function of the private sector may actually serve to provide them with an added source for grievance against the Soviet state. On this count as well, the current reevaluation of the role of the private sector may thus turn out to be counterproductive.

In conclusion, the really crucial question to be asked is of course *why* private activities suddenly should have ceased to give rise to all those problems that have been outlined above? For this Vasilev provides no explanation, and the same holds for the recently made statements by Nikiforov, Kalinkin, Kuznetsova and others. This fact certainly leaves the field open for speculation along the lines of wishful thinking or ideological window dressing, in support of the private sector. Although the effects may be positive in the short run, the current wager on the plots may over time turn out to have quite an important backlash effect, of a rather unpleasant nature. An examination of the latter shall be our concern in the concluding chapter of this study. Before proceeding, let us summarize briefly the main ingredients in our composite picture of a peasant Loyalty – Soviet style.

A Passive Loyalty

As we have indicated above, in Hirschman's terms Loyalty was a mechanism that served to delay Exit and thus to activate Voice, which in turn would promote recuperation. This we shall refer to as an *active* Loyalty and its attainment will be seen as a necessary condition for success along that path of *perestroika* and economic reform, which has been embarked upon by the Gorbachev leadership. In the actual case that has been described above, however, we have a somewhat different process, the nature of which we shall illustrate by decomposing the function of Hirschman's original concept of Loyalty into two steps. The first of these will represent the delay in Exit and the second the activation of Voice.

The function of delaying Exit is obviously achieved in both the present and the original Hirschman case, albeit in radically different manners. By introducing the concept of a 'soft' Exit, we have gone beyond the simple dichotomy originally used by Hirschman. This has two important implications. On the one hand, the restrictions that exist against a 'hard' Exit will no doubt serve to produce the delay. On the other hand, the 'soft options' will simultaneously serve to defuse Voice and thus to block the desired stimulus for recuperation. In essence, this can be presented as a strategy of neither Exit nor Voice – or of Stay-Silence – which rests on the practice of the various soft Exits and on the associated consequences.[47] What is achieved by ways of these mechanisms is something that we shall refer to below as a *passive* Loyalty.

The important question to ask, in this context, is what the implications are of the distinction that is thus made between our perspective and that of Hirschman. In order to answer that question, we shall return briefly to the previous discussion of the concept of Loyalty as such. Our distinction between an active and a passive version reflects the two very different functions that can be filled by Loyalty, in an instrumental and perhaps rather subtle interpretation. The *active* version is connected with a sense of being incorporated into and identifying with the system. The corresponding state of mind will consequently be such as to induce the individual to undertake certain actions with the aim of helping to improve that system. This relates to the previously mentioned process of a coherence between *agitprop* and perceived reality. In essence, it represents the original interpretation of the function of Loyalty.

The *passive* version, on the other hand, is connected with our various soft Exits, i.e. with a feeling of being alienated from the system. The corresponding state of mind will in this case be such as to induce the individual to refrain from any actions designed to improve the system, and maybe even to undertake actions which actively seek to damage it. This latter version of Loyalty embraces that negative process which is connected with a growing discrepancy between utopia and reality and forms the real mainstay of our mechanism of political stabilization. Let us proceed now to look more closely at the respective costs and benefits involved.

Stability First

The main requirement of a policy that derives from a desire for control is of course to block a 'hard' Exit. As we know this was achieved early on, via the forceful introduction of the *kolkhoz* system. After this initial step, however, it must soon have become painfully obvious to the Stalin leadership that simple repression would not suffice in order to contain such discontent and frustration that followed in the steps of compulsion, at least not if any hopes of continued agricultural production were to be maintained. Something else was needed and it is the ingredients in the response to that realization which would combine to form our mechanism of passive Loyalty.

As Hirschman has pointed out, there is an inverse relation between Exit and Voice, such that pressure on the latter will increase if the former is suppressed, and vice versa. A case in point – which is used by Hirschman – is Castro's Cuba, where an important initial move, after the revolution, was to allow an Exit (to Miami) for those who might later have become vociferous critics of the system. Thus, the general level of Voice in the system was reduced, and less internal pressure resulted when Exit was eventually curtailed as well.[48] Soviet mass collectivization, however, allowed for no such initial depressurization, as the aims were set for a 'full' (*sploshnaya*) collectivization to be achieved in a brief period of time. Consequently, pressure inside the system grew dangerously high and in order to prevent a total collapse of food supply a number of soft Exits had to be allowed. Trade on the *kolkhoz* markets and rights to private plots are the most obvious examples. Moreover, they are also the preferred ones, since they represent what we have referred to above as *productive* soft Exits.

Over time, however, these initial forms of soft Exit proved insufficient to compensate for the growing discontent, or – in Hirschman's terms – for the continuing process of 'decline'. The status of the *kolkhoznik* was serf-like to say the least. Via the system of domestic passports, he was tied to the land, and frequently the only pay for his work in the socialized sector was the right to a private plot. If we then add in the various obstacles erected and the harassment undertaken by local officials against the productive Exits of plot production and market trade, we have the makings of a process that is sometimes referred to as a degeneration of 'social tissue'.

This latter process can be interpreted as a gradual increase in the price that had to be paid in order to maintain political and social stability. The system was passively and perhaps unwittingly forced into accepting and accommodating a different form of soft Exit, an Exit which featured drinking, slacking and general apathy, and which we have referred to above as a soft *improductive* Exit. The reality that underlies this form of Exit has of course been observed not only by Vasilev. There is plenty of other similar evidence, some of which is rather striking. Suffice it here perhaps to mention Lev Timofeev's book on *The Peasant Art of Starving* and the ethnographical study of a Siberian *kolkhoz* – the *Karl Marx Collective* – made by the British anthropologist Caroline Humphrey.[49]

Such Exits, moreover, represent a continuous process, rather than a distinct historical phase. As more and more resources are being poured in, to fill the gaps left by the various soft Exits, the process of degeneration of the social tissue is accelerated. The peasants experience that both productive and improductive forms of soft Exit are permitted, albeit grudgingly. The price of stability will thereby gradually increase, as yet further resources have to be poured in to compensate. We have here the very essence of that process which underlies our metaphor of likening Soviet agriculture to a Black Hole. Although its detrimental effects on the system as a whole make the improductive form of Exit different from the productive one of plot or market activities, we should note that the two also share an important trait, in acting as a safety valve that serves to defuse Voice – constructive as well as disruptive – against the system.

Hence a common denominator of the components in our answer is that the system is provided with a basic political and social stability, at the cost of a gradual erosion of economic efficiency and of vital feedback mechanisms regarding the performance of the system as a whole. This brings us over to the second dimension of our answer, that concerning the nature of the costs that are incurred in the process.

A Costly Victory?

It is certainly trivial to note those short-term effects of reduced efficiency and of a fragmentation of production that follow from the practice of the various soft Exits and we shall not waste any further time in discussing them here. Instead, the remainder of our study will examine in somewhat more detail the longer-term problem of the moral and psychological impact that they may have on the peasants. Before doing so, however, we might emphasize at this point that if we think purely in terms of the perspective of control *per se*, that we have seen repeatedly used by Yanov, the achievements of those acts and measures which have combined to produce our mechanism of passive Loyalty must be deemed as quite successful. It may be illustrative here to mention H. G. Wells' highly visual account of a personal meeting with Lenin in the Kremlin, in 1920: 'At the mention of the peasant, Lenin's head came nearer mine, his manner became confidential, as if after all, the peasant might overhear.'[50] In a similar vein, we can quote Churchill's recollection of a conversation with Stalin, on the ordeals of mass collectivization: 'It was terrible. Terrible. And four years it lasted.'[51]

In short, it has proved possible to eliminate from Soviet society the troublesome 'peasant' class, which in itself is certainly no mean achievement. From the belligerent class that was responsible for the uprising of 1905, for the revolution in February 1917, and for once almost breaking the spirit of the great Stalin himself, the Soviet peasantry has been transformed into a 'Loyal' mass of atomized individuals who, depending on the form of soft Exit, are either busy making souvenirs for tourists or growing flower seeds for the market or are deeply submerged into a state of drunken apathy with little interest for the activities of the Soviet state or the Communist Party. This portrayal is most certainly an extreme one, but it does illustrate that the path of development hardly leads in the direction of creating a modern industrialized farm sector. Whether this policy has actually been a rational one, or if the outcome has been worth the costs would thus seem to be largely a matter of preference. These, however, are questions that shall be left for the concluding chapter.

Conclusion

In the present chapter we have swung the pendulum of our presentation firmly over to the peasant end of the spectrum. From the account of actual developments and formal rules, we have gone via a discussion of the Soviet debate on the matter, to various practical manifestations of negative attitudes to private activities. One important impression that can be gained from the presentation thus far is that of an implicit shift of responsibilities, from an inefficient and in various ways demoralized socialized agricultural sector onto the shoulders of various other initiatives. Responsibilities in production are gradually shifted from the *kolkhoz* and *sovkhoz* system, to the private and official fringes, whereas responsibilities in marketing are being moved from the official system of procurement and distribution, to the *kolkhoz* markets and to those different arrangements that we have referred to as 'structural autarchy'.

Viewed in purely economic terms, these shifts are obviously not yet of any major importance and it is certainly not the case that the socialized systems of production and distribution are in a process of being dismantled. If, however, we look at them from the perspective of eroding economic responsibility and of accommodating various inefficiencies that derive from peculiarities in the overall system of planning and control, a somewhat more serious picture will emerge.

Against this background, it is hardly surprising that the current policy of *glasnost*, and of 're-evaluating' the role of the private sector, has produced arguments which are aimed quite simply at dissolving 'backward' *kolkhozy* and *sovkhozy*. In a recent issue of *Pravda*, for example, Shmelev has stated that 'in a number of cases one might even liquidate *kolkhozy* and *sovkhozy*'.[52] In a recent book, he has expanded on that rather controversial statement. In response to the hypothetical but rather realistic question 'What? Then there will not be any *kolkhoz* anymore?', the following statement is made:

> Well, what should one answer? Sure, it is true that there will be no *kolkhoz*. Where it is not suitable, there is no place for it. After all, the word *kolkhoz* is no more than an acronym to which we have grown accustomed, and this must certainly not mean that it is the one possible and untouchable form. Briefly put, where there will not be any *kolkhoz* there will instead be another form of organization of production, equally cooperative, equally socialist.[53]

It may be of some importance to note that the context of Shmelev's statement is that of advocating family contracts and support for the private sector, rather than a full liquidation of socialist agriculture. Nevertheless, this certainly does not diminish the heresy in questioning the *kolkhoz* and the *sovkhoz* as the basic forms of organization of Soviet agriculture.

We have shown above how a desire for political stability has served to depeasantize Soviet agriculture, with the effect of demoralizing the peasantry as producers of food for the population, and of placing a seemingly ever increasing burden on industry – for labour, on the state budget – for investment and subsidies, and on scarce foreign exchange holdings – for imports. Has it all been worthwile? Are the processes reversible? To these questions we shall now turn, in the final and concluding chapter of the study.

Chapter seven

A Pyrrhic victory and its consequences

In this concluding chapter, we shall argue that the achievements of the private sector in Soviet agriculture – and of the two supportive fringes – can insome respects be considered as a 'Pyrrhic' victory. On the one hand, it is no doubt the case that those achievements which have been registered to date represent genuine benefits, in the sense that valuable contributions have been made in areas where the shortcomings of the socialized sector have long been notorious, i.e. in the production and distribution of foodstuffs. Without the plots and the *kolkhoz* markets, an already difficult food situation might well have deteriorated beyond the tolerable. In accordance with Soviet parlance, we have referred to this role as the 'support' function of the private sector.

As we have repeatedly indicated in the preceding chapters, however, there is also an alternative and somewhat less flattering interpretation of this function, an explanation which implies that in a longer-run perspective it may well transpire that the support in question has been bought at a price which is substantial in magnitude and perhaps also rather unexpected in nature. This explanation focuses on the formation of attitudes and the conditioning of human behaviour, and it is of course in some sense a rather speculative one. It does, however, place the human element at the centre of attention, and thus agrees well with Gorbachev's statement to the 27th Party Congress, in February 1986, that 'the main driving force in the advancement of the agro-industrial complex, indeed its very soul, is, has been and will remain man.'[1] It is from this perspective that we shall argue that there are some broader implications of the current wager on the plots which may not only escape a full appreciation by Soviet policy makers, but may perhaps even serve to strengthen existing obstacles to the possibility of eventually reforming the system.

The time has now come to substantiate these rather far-reaching claims. This will be done in two stages. First, we shall complete that discussion of the moral and psychological dimension of plot activities which was begun, in the previous chapter, by looking at the situation of the individual at the micro level. Here we shall shift the discussion onto the macro level, by concentrating on the role of ideology in a public policy which is officially aimed at influencing peasant attitudes in the direction of creating the elusive Soviet Man. Our second step will then be to turn the presentation 180 degrees around, in order to examine the requirements for success in making the transition to the second part of the original Hirschman process of promoting recuperation, i.e. that of activating Voice. Here we shall return to the distinctions made above between a *passive* and an *active* Loyalty, and to the claim that the two represent social processes of such fundamentally different natures that taking the step from one to the other may be no simple task.

Voice and Opposition

Agriculture versus Industry

Before proceeding to our argument proper, however, we shall digress briefly into the realm of opposition to the regime. This will be done with the explicit intention of claiming that there is in this respect a qualitative difference between agriculture and industry, a difference which in a long-run perspective places the Party's policy towards the peasantry in a quite distinct category of its own. The first step in this argument will be to underline that what we understand by 'political opposition' is something considerably different from the normal Western understanding of that concept.

In a setting such as the Soviet, which permits no openly organized opposition, we are perhaps better off talking instead about anti-systemic sentiments. The importance of this distinction lies in bringing out that potential threats against the regime may develop just as easily out of alienation and inaction, as out of directly hostile activities. We would thus argue that Silence and soft Exits may be just as dangerous as openly critical Voice. The difference simply lies in the time perspective. While the latter may certainly be de-stabilizing in the short run, the former may in the long run serve to produce serious economic and social dystrophy. It is in this context that the distinction between agriculture and industry shall enter the stage.

It is of course true, as we have indicated above, that much of that frustration and discontent which we have ascribed to the peasantry may be found among other segments of the population as well. Below, however, we shall argue that the implications in the case of agriculture are of a more serious nature. This argument rests on the assumption that industry does not have any 'private workshops' corresponding to agriculture's private plots. One parallel which might be advanced from the industrial side is of course the second economy, but this is a problematic one. As these activities are largely illegal, they will not act to form an openly accepted alternative structure in Soviet society, as do the plots and the markets in agriculture. Another potential parallel from industry concerns those cooperatives which were abolished in the 1960s. Some say that this decision has increasingly become a subject of regret among Soviet policy makers, and in this light it is interesting to see a new law on cooperative activity being recently adopted. We shall return to this in a moment.

The main reason for underlining the existence of a difference between the two sectors in this respect, concerns the fact that agriculture's plots and markets combine to form an alternative social and economic structure, largely outside Party control, an institutional structure which serves as a *locus* for Exit from the socialized sphere. The simple fact that there is somewhere else to go, some other object on which to expend time and effort, will have two rather serious implications for the prospects of eventually socializing the peasantry into the Soviet system and its ideology. First, a *focus* will be provided for the emergence of a set of values and beliefs that are seriously out of tune with the official ones. Those negative attitudes and hostile actions which have been presented above will serve here to provide precisely that 'ray of sunshine' which was seen by Vasilev to promote the growth of that side of peasant psychology which is connected with 'private property instincts'. Second, as was seen above, even if the peasants were actually prepared to absorb the official message in the first place, which is certainly a debatable proposition, they will quite simply be too busy to be able to do so.

We certainly do not intend to imply that there is a simple dichotomy between agriculture and industry, as far as the options for soft Exit are concerned. This is obviously not the case. Nevertheless, we shall argue that there are two important differences of degree which do serve to place agriculture in a separate category. First of all, it can be argued that the Exit available to frustrated workers is predominantly that of a soft improductive retreat to within oneself, via drinking, apathy and cynicism. This pattern of response certainly has its obvious parallel in agriculture, but if we assume that the share of the rural population that is engaged in plot and market activities is larger than that of the urban population which is active in the second economy, we may conclude that soft Exit in agriculture will to a greater extent be of the productive kind. In a short-run perspective, this would no doubt seem to place agriculture in an economically more advantageous position.[2]

In a longer-run perspective, however, we must also consider the social effects of the soft productive Exit. Here the difference between the two sectors concerns atomization versus incorporation. In contrast to plots and markets, as we have argued above, the second economy does not present an alternative social and economic structure. It is not only formally illegal, and thus associated with danger and risk. It also lacks accepted physical and geographical locations where large groups of people can meet and exchange views on common everyday problems in relation to the nature of their Exit. While the peasants will thus to some extent be incorporated into an openly tolerated alternative structure in Soviet society, the urban population will carry out its soft productive Exit mainly under cover and in isolation from each other. Here we must say a few words about the new law on cooperatives.

The Law on Cooperatives

In the fall of 1987, guidelines for allowing voluntary cooperative activity were issued and in March 1988 a draft law was published.[3] In a broad context, this law is obviously related to the 1987 law on 'individual labour activity' and to the 1988 model *kolkhoz* charter, both of which have been discussed above. Its relation to the former lies in its boldness in allowing the employment of wage labour and in not explicitly stating that cooperative membership must not be a main form of labour activity. Where 'individual labour activity' could be no more than a soft Exit, cooperative activity may thus actually come close to a 'hard' Exit, allowing able-bodied labour to leave the state-controlled sphere altogether. This considerably more liberal stance, which even goes so far as to provide for foreign trading rights and joint ventures with capitalist firms, reflects rather clearly that the notion of 'cooperative' is much more palatable to Soviet ideologues than that of 'individual'. So far, the Soviet case exhibits important parallels with the highly successful story of cooperation in Hungary. This apparent liberalism, however, is largely restricted to paper.

In practice, the new cooperatives will depend on local authorities (formally the local Soviets) for licensing and for 'material supply', and they will be subjected to an unspecified but differentiated and progressive taxation. In this light, it is interesting to note that provisions are made for establishing cooperatives in association with (*pri*) state enterprises, in which case they could draw on the latter's supply sources. Cooperatives are also explicitly permitted to compete for the new 'state orders' (*goszakazy*) which would be another way of securing inputs, in the absence of a market for these purposes. With sufficient determination amongst top policy makers it would no doubt be possible gradually to overcome the logistical problems. Of greater importance, however, is that – in stark contrast to Hungarian experience – there seems to be little enthusiasm among the Soviet population for such cooperative undertakings. The following is Hanson's comment:

> Besides this, there is widespread popular resistance to the idea of cooperatives. Many Russians believe there is no such thing as socially useful enterprise – only theft. Writers of letters to the press speak of the recent measures on cooperatives, individual labor activity, and family contracts as retrograde and refer to 'the rebirth of the kulak'.[4]

As far as the urban population goes, one would thus want to excercise some restraint in expectations for the future. There will no doubt be a considerable interest in forming cooperatives as simple fronts for illegal second economy activities, but this the authorities will be looking for and be ready to stop if possible. The controls that will inevitably be associated with the latter desire add further scepticism regarding the autonomous status of the urban cooperatives. What then about the rural population?

The relevance of the new law on cooperatives to the new model *kolkhoz* charter lies quite simply in the fact that the two form rather obvious contradictions to each other. Formally, the *kolkhoz* already constitutes a voluntary cooperative of peasants, and one certainly wonders how a new set of voluntary agricultural cooperatives can be slotted into this picture. With the 1988 model charter explicitly allowing peasants to leave the *kolkhoz*, one might perhaps think that the 'new' cooperatives should crowd out the 'old', since the former will most likely be free of bureaucrats and other such 'parasites', who are currently under heavy attack in the official rhetoric. This, however, will certainly not be allowed to happen, and one may thus be justified in questioning the true intentions of the new attitude towards cooperation. Perhaps one might also wonder if the peasants would really be interested in giving up the security of the socialized sector, in order to venture out into the unknown. We shall return to this latter aspect in a moment.

Most likely, the new cooperatives will find a place alongside the existing structure, thus becoming a further addition to the already peculiar hybrid of large-scale socialized farms and tiny private plots. They can, however, be expected to play a distinctly new role. While the cooperatives are supposed to be guided by a profit motive, and to choose for themselves inputs, outputs, employment and wages, the *kolkhoz* will continue to work under state procurement plans, perhaps relabelled as *goszakazy*, and will no doubt also continue to be subjected to that extensive form of 'petty tutelage' which is so well known from the past. The cooperatives, however, are relegated to holders of private plots, who may join together for a variety of purposes, even to hire outside contractors. The importance of the private sphere may thus increase, but only within certain limits. We may recall here that membership in the *kolkhoz*, and fulfilment of established work norms, still remains a precondition for being entitled to a private plot in the first place. We thus have a maze of horizontal and vertical lines of association and subordination, the outcome of which hardly inspires much hope. The crazy organizational construct that is represented by this *kolkhoz-cum*-private plots-*cum*-family contracts-*cum*-new cooperatives could only have sprung out of the Soviet bureaucratic machinery. Let us return now to our main track.

A Polish Parallel

Given the central role that will be played below by the various mechanisms of incorporation, we might profitably draw a parallel here with recent developments in Poland. It is sometimes argued that the popular response to repeated crises and regime failures has been that of an 'internal emigration'. This refers to the emergence of an *alternative* Polish society, marked by ideals and values that are very different from those of the official sphere, a society which features flying universities, underground Solidarity, a highly active output of *samizdat* and, of course, the Catholic Church.

The role that has been played by the Church in Poland is precisely that of providing a *locus* for Exit, i.e. somewhere to go, which was outlined above. It has provided the physical places to meet, under auspices tolerated by the regime but largely outside the control of the Party. In that process it has also – albeit passively so – provided a *focus* for the articulation and aggregation of Voice, i.e. opposition, against the regime.[5] On the one hand, we would certainly agree with Roman Laba, in pointing out that the real driving force of opposition in Poland has been a strong groundswell of unfulfilled worker demands, which have been 'left over' from previous occasions of unrest, and that KOR and the Church have appeared 'in the role of helpful and intermittently heroic auxiliaries rather than initiators of Solidarity'.[6] On the other hand, however, we would also argue that in the absence of that institutional structure which has been provided by the Church, there would have been nothing to capture the groundswell of worker demands. Hence none of the events that have taken place since December 13, 1980 might ever have come about.

We shall argue below that a similar function is filled in the Soviet case by the plots and the *kolkhoz* markets. This parallel is no doubt weak in the sense that the level of political culture in the two societies is considerably different. It is, for example, hardly realistic to expect that Soviet peasants, in an otherwise similar situation, would undertake actions akin to those of Polish workers. The issue at hand, however, is one of principle, concerning structures and the possibility for action, rather than actual content. In essence, we are dealing here with the ability of the authorities to control Voice against the regime.

Controlling Voice

To develop this theme, we shall borrow an interesting framework from a recent book by Alexander Motyl. The theme of the book is the threat to the stability of the Soviet system that is inherent in the nationalities issue and Motyl suggests to investigate his problem by looking at the activities of individuals in Soviet society in a two-dimensional perspective of time and space. This perspective produces a number of 'boxes' which contain different types of activities and are subject to different types of control. Time spent in private, within the confines of the home and the family, is seen to be largely free of control and thus to form a safety valve which roughly corresponds to the function of our various soft options. Time spent in public is, on the other hand, divided into collective activities, which are placed firmly under Party control, and private activities, which are subject to control by the KGB.[7]

The main point of this presentation is the recognition of the private time-space as a form of an asylum. This not only means that the KGB, in stark contrast to the Stalin period, has been 'relegated' to one box only. It also 'relegates' the outlet of frustration and discontent to the private homes of the citizens, a dimension where it can do comparatively little harm. In this perspective, the chief function of the KGB becomes to maintain that dividing line, to make sure that any individual who ventures into public space also leaves behind any potentially deviating beliefs.

The relevance of Motyl's model for our presentation lies in its rather clear illustration of how the private sector in agriculture has come to straddle precisely that vital grey zone between the private and the public spheres which the KGB is supposed to keep clean. Activities on the plots are not subject to any formal control. Nor is the process of taking produce to market. Moreover, with the recent abolition of the requirement for documents of ownership in order to trade on the markets, even these activities are placed largely outside formal control. Those various forms of harassment that have been described above are probably better seen as an illustration and a manifestation of this absence of control than as a substitute for it.

Motyl clearly recognizes the importance of this phenomenon, and places particular stress on Soviet tendencies to form new 'non-autonomous' (i.e. Party controlled) organizations as soon as a new popular interest or activity is perceived to be growing, the most absurd example being a quoted suggestion by some Ukrainian Party officials to turn dance halls into 'political discoteques'.[8] This presentation is strikingly reminiscent of White's 'mechanisms of adaptation'. In our own framework we would characterize such organizational activities as an ambition to capture and channel Voice by incorporating individuals into the socialized sphere of society.

Since the concept of Voice will have a crucial role to play in our discussion below, a few points of definition might be in order at this juncture. First, since it cannot be articulated and aggregated autonomously, as is the case with opinion and pressure groups in the West, it does of course not represent Voice in the original Hirschman sense. What is involved here is instead something that we have referred to elsewhere as 'soft' Voice.[9] Since it does in a limited sense represent human communication, it fills the dual function of (a) providing the individual with a safety valve, and (b) providing the regime with some information feedback. The former function is similar to that of the soft Exit, and thus merits no further comment.

The latter, however, represents the real crux of the matter. On the one hand, it is obvious that the regime needs in some way to generate crucial feedback about the performance of the system at large, and preferably in the process also to instill into the citizens a feeling of participation. This is the essence of White's mechanisms of adaptation and incorporation, and it calls for leaving open at least some channels for communication from below. On the other hand, however, it must also prevent such communication from assuming a dangerous *form*, such as factions or interest groups, or a dangerous *content*, such as demands, for example, for free elections or open borders. This is the essence of Motyl's presentation of 'non-autonomous' organizations, and it places the need to control Voice in focus.

The relevance of the Polish example concerns precisely those dangers that are inherent in allowing an alternative and truly autonomous structure to emerge, a structure which may attract Voice and allow it to assume both the prohibited form and the prohibited content. The main ingredients of the Polish story should be well known, and will not need repetition. Instead, we shall argue that the contrast between agriculture and industry, which we have just made in the Soviet case, has a peculiar resemblance to the Polish phenomenon of 'internal emigration'.

The private sector in Soviet agriculture is conspicuous precisely by the relative insignificance of any attempts to 'incorporate' the peasants into the official sphere of Soviet society. One might perhaps even convincingly argue that there has instead been a policy of ostracism, if not consciously and deliberately, then at least in practice.

Symbolically, this may be represented in Mukhina's 1937 statue 'Worker and *Kolkhoz* Woman', which stands outside the VDNKh exhibition in Moscow and which is also the official logo of Mosfilm. The male figure symbolizes industry and the growing strength of Soviet society, while the female represents the soil and 'Mother Russia', leaving the peasants little scope for identification and incorporation. The great stress that we have seen placed on the organized nature of the private fringe, on the creation of gardening associations and of TTOs for the *dachniki*, illustrates the importance of the problem at hand but evidence indicates rather limited success.

It is the resultant strategy of Stay-Silence, i.e. of formally remaining within the official structure while producing no Voice that can be captured by channels belonging to that sphere which we have sought to illustrate in our mechanism of a passive Loyalty. The benefit is political and social stability, the cost a loss of initiative and feedback. Below we shall argue that the social processes that underlie this peculiar form of 'Loyalty' may be difficult – indeed perhaps impossible – to reverse in the near future. Before proceeding to that argument, however, we shall complete our discussion of peasant mentality, by looking at what tools may be available to Soviet policy makers for capturing the peasantry.

Ideology and Rationality

The Initial Problem

In order perhaps better to understand the tasks that have been placed before Soviet ideologists and policy makers, striving to lift the reality of agriculture up to the level of the promised utopia, we may begin by recalling from above the nature of the really central problem at hand. In quite simple terms, we can illustrate this with an example that has been provided by Yanov. The setting is the attempts that were made during the 1960s to introduce into agriculture a system of self-determination and pay by results, based on 'normless teams' (*beznaryadnye zvenya*), and sometimes known quite simply as the 'link' movement.[10] The theme of the example is an engineer at a machinery testing station who, having developed some new equipment that showed highly promising results on the experimental fields, was puzzled by the total lack of success in the *kolkhoz* fields. When visiting one of the farms in question, however, he soon enough found the reason. The new equipment required precise handling and could not be driven at speeds over 2.5 km per hour. The tractors, however, were capable of doing 24 km, and as soon as no supervisors were in sight that was indeed the speed that they would do.[11]

A Pyrrhic victory and its consequences

This example captures the real essence of that process which we have described above as presenting Soviet agriculture with the image of a Black Hole. Seen from the point of view of the collective, shallow ploughing is quite obviously absurd, as it will have a seriously detrimental influence on sowing, and thus on the eventual size of the harvest. As a group, all will be worse off. Seen from the point of view of the single individual, however, we face an altogether different logic. Since the work assignment of the tractor driver is measured in hectares, it is quite rational for him to plough shallow. This will increase his speed, which allows him to cover more hectares and thus to earn a bonus. Moreover, by not engaging the plough to its required depth he will not only conserve fuel but also reduce wear and tear, both of which will earn him additional bonuses.

What we have here is an illustration of what is known in Game Theory as a Prisoner's Dilemma situation. Each single individual pursues a strategy that from his point of view is quite rational, but the outcome of the actions of all individuals taken together will be irrational, since all would have been better off by choosing an individually less preferable initial strategy.[12] Soviet agriculture is rife with situations of this nature, where a collectively rational outcome is blocked by the nature of incentives facing the single individuals. It is this which has prompted Vasilev to call for a 'return to the soil', and to bemoan the absence of a link between effort and reward for the individual peasants, and it is this which has prompted Yanov to talk about the 'depeasantization' of Soviet agriculture. It is, however, not something that is found exclusively in either agriculture or in the Soviet system. By underlining this important fact, we shall perhaps be better prepared to understand why the consequences of such situations are more serious precisely in agriculture and in the Soviet system. To do so we shall make a short digression into the realm of ideology and morality.

The Role of Ideology

In his above-mentioned book, Motyl makes the important observation that the 'ideologies of democratic states differ from those of authoritarian ones with respect only to content, not to function', and it is certainly true, as he goes on to note, that while 'American idealization of George Washington may not be so effusive as Soviet exaltation of Lenin', it certainly serves the same purpose: 'to provide a time-honoured, glorious reference point for the mass of citizens being socialized in the present.'[13]

If we thus start from the two premises that (a) ideology is nothing specific to the Soviet system, and (b) that ideology is more 'visible' in the Soviet system, we are still left with the question of the general function of ideology. Here the American economic historian Douglass North has provided an interesting approach, by pointing out a dilemma that is built into the neoclassical economic model. On the one hand, this model assumes that all individuals act so as to maximize their own personal welfare, but on the other, they are also assumed to be able to agree on a given set of rules for this behaviour, the latter being an obvious prerequisite for a viable political system. Rational action on the first count, however, necessarily implies irrationality on the second. As North points out, it is 'in the interests of the neoclassical actor to disobey those rules whenever an individualistic calculus of benefits and costs dictates such action.'[14]

At the root of this problem lies the Prisoner's Dilemma, as outlined above. We would probably all agree that littering, for example, leads to a reduction in total welfare. For the single individual, however, there is a cost involved – in terms of time and effort – in *not* littering, while the impact on the environment of *his* littering is negligible. Since his loss of welfare on the first count most likely exceeds that on the second, the rational action would clearly be to litter. North's pointed question is: 'How much additional cost will I bear before I become a free rider and throw the beer cans out the car window?'[15]

In this formulation, the answer is obvious. If all individuals act rationally we are clearly destined for the Prisoner's Dilemma. If nobody would litter, all would be better off, but given the rules of the game this outcome is blocked – the attraction to the individual of littering is simply too great. Enforcement would be one possible way out, but for most rules underlying a working society the costs of enforcement would most likely be greater than the increase in welfare that could be thus achieved. It is obvious that it pays to enforce rules against, say, murder, but what about jaywalking, or swearing in public? If all individuals were to act rationally, from their own individual points of view, we would expect very few of the rules that constrain their behaviour to be obeyed and thus society should break down. Yet, we observe that most people abide by most of the rules most of the time, although it is costly for them to do so. It is here that ideology enters the picture.

The root of the evil in the free rider problem is the attraction to the individual of attempting to increase his own personal welfare at the expense of others. In order to solve this problem we must increase the price that he has to pay for doing so and, as we have noted above, legal and economic sanctions might not be a feasible way of achieving this. In such cases we are left only with imposing *moral* costs. By instilling certain norms and values into people, ideology may help simplify their decision-making process – as certain decisions will be 'out' – and thus also help socialize them in the given social environment. Ideology becomes a cost-saving device in the building of a society. As North puts it: 'Strong moral and ethical codes of a society are the cement of social stability which makes an economic system viable.'[16]

With regard to the mechanisms that are at play here, North also makes two further points that are of direct relevance to our presentation. First, he notes that the 'costs of maintaining the existing order are inversely related to the perceived legitimacy of the existing system.'[17] This implies that the larger the gap between the beliefs and values that are held by the population at large, and those of the official ideology that is propagated by the regime, the harder it will be to sustain the latter. The case of Soviet agricultural policy, from 1917 onwards, would seem to be a text-book illustration of this rather obvious principle, where the outcome has been a failure to close the gap. North's second point concerns the dynamics of the latter problem:

> Individuals alter their ideological perspectives when their experiences are inconsistent with their ideology. In effect, they attempt to develop a new set of rationalizations that are a better 'fit' with their experiences.[18]

This latter process is precisely that which we have tried to capture above, in our mechanism of a passive Loyalty. It illustrates the importance of Elster's discussion regarding 'states that are by-products', and it is clearly relevant for those references to a dual nature of peasant psychology, which we have seen made by Shmelev and Vasilev. If people repeatedly experience that actual reality has a poor correspondence with the utopia that is presented by the regime, the credibility of the latter will naturally suffer. Soviet history in general, and that of Soviet agriculture in particular, provides an ominous illustration of how the process of alignment between official ideology and widely held beliefs and values may break down. The case of the peasantry is special due precisely to the 'dual nature of psychology', which in turn is due to the existence of that 'alternative structure' which we have argued consists of the plots and the *kolkhoz* markets. Since this is of crucial importance to our argument, let us subject that case to a somewhat closer scrutiny.

The Making of a 'Private' Peasantry

As we have seen on repeated occasions in the preceding chapters, the private sector is frequently assigned an important 'educational' role. We have seen this expressed above by academician Nikonov, by Shmelev and by others. Here we may add the voice of V. I. Sidorenko, another protagonist of activities on the plots, who takes a similar stand: 'We must not forget the social factors: a direct attachment to the land, reduced labour turnover, a more active formation of work teams, decreased migration to the cities, increased rural labour supply.'[19]

Given what has just been said about the role of ideology in shaping mentality and in subconsciously preventing people from taking 'anti-social' actions, this role of private agricultural activities is naturally placed in focus. The question, however, is in which direction the current wager on private initiative will act to influence these variables, i.e. which states of mind will result as by-products of the direct action. In order to understand the present, it might perhaps be a rather good idea to start by looking at the past. The following is Lewin's verdict over the outcome of the first chapter of *kolkhoz* history:

> In other words, the social effects of the functioning of the *kolkhoz* system, for a generation at least, consisted in reproducing backward Russian *muzhiki* instead of the modern cooperative, industrialized farmers. If anything, these *muzhiki* might have become more interested in private property and private farming than they ever were before.[20]

Lewin argues that while the concept of private property was very weakly – if at all – developed in Russia before 1906, 'collectivization, although aiming at uprooting such attitudes, went in fact a long way toward reinforcing and developing them.'[21] The message that is contained in this statement is rather astonishing. Not only was the individualistic side of peasant mentality allowed to grow, it was perhaps even the case that previously non-existent concepts of a rather alien nature were introduced into the picture. In this perspective, the first stage of Soviet agricultural policy must certainly be understood as having been clearly counterproductive. Instead of breaking up the old peasant community, the *mir*, and transforming the peasants into collectively minded agricultural workers-*cum*-Soviet Men, it succeeded in destroying whatever collective sentiments once existed in the *mir*, substituting instead true *muzhiki* with little interest in either progress or the values of the Soviet state.[22]

There lies in this process the makings of an irony of almost monumental historical significance. The expression 'wager on the plots' has been consciously used above in order to bring recollections of that 'wager on the strong', which was once the slogan of Premier Stolypin's attempt, following the abortive 1905 uprising, to promote a capitalist transformation of Russian agriculture: 'The Government has placed its wager, not on the needy and drunken, but on the sturdy and strong.'[23] As we know, the experiment was cut short by the assassination of Stolypin and by the 1917 revolution, but it is important to note the belief of some writers, that even if it had been granted enough time, the result might still have been disappointing.[24] The irony lies in the fact that where Stolypin failed to break up the *mir* by promoting capitalism, Stalin succeeded by promoting socialism.

Forced collectivization is of course a matter entirely different from the current encouragement of plot activities, but so is the initial position, or the point of reference. The really crucial factor at play here can be illustrated in Vasilev's and Shmelev's repeatedly used picture of peasant psychology as having two sides, one 'collective' and one 'individual'. The origin of this mental bifurcation is the economic and social duality that is represented by the existence of private plots inside the collective farms. As Lewin puts it, the private plot 'had a paradoxically disproportionate influence: it was an economically restricted but socially powerful force in the shaping of a class.'[25] Over time, we shall argue, the policy of grudging acceptance and periodic harassment that has been described above has served to reinforce and perhaps even make permanent this duality. We may recall, for example, Vasilev's claim that the introduction of a guaranteed element of pay had led to an unfortunate disconnection of the tiller from the soil, to a growth of that side of psychology which is connected with 'private property instincts'.

When Stalin set out to collectivize the peasants they may have had little idea of 'private property instincts', but in the process of being forcefully turned into worker-collectivists they were certainly 'infected' with precisely that mentality. When Gorbachev – and his immediate predecessors – have set out to promote the private sector, the duality has already been there. Encouragement of private activities can thus be assumed to have served to reinforce the 'private' rather than the collective side. This would certainly remain true today, irrespective of what the proponents of the new policy may have to say about the allegedly organic harmony between the two sectors.

The dangers that are inherent in this process have of course been observed by Soviet writers as well. Vasilev, for example, surveys and comments upon a number of letters, published in Soviet newspapers, which contain warnings about the serious risk that the current policy of support for the private sector may actually lead to the reemergence of the hated *kulaks*, the class of better-off peasants which was once so successfully uprooted by Stalin.[26] It is of course possible to brush off such dangers by using the type of argument that we have seen repeatedly displayed above, i.e. the overall conditions of Soviet agriculture, and the nature of production in the private sector, are such that no 'private' peasants can reappear. Terminology apart, however, it is fairly clear that there is a risk of spillover effects and that some people at least take it seriously. If we were to give negative answers to all of those largely rhetorical questions about the allegedly 'socialist' nature of the private plots that have been posed above, by Shmelev and others, then the prospects for the future of Soviet socialized agriculture look rather bleak.

The Creation of an Active Loyalty

If Gorbachev is to succeed in his attempt at transforming Soviet agriculture from a Black Hole into a modern efficient sector of the economy, ready to serve as that engine of growth that it was once intended to be, he will need precisely that which we have referred to above as an *active* Loyalty, i.e. that people delay Exit and resort to Voice in a sincerely felt hope that this will serve to promote recuperation. At all levels of society, i.e. from the top bureaucrat and down to the junior milkmaid, he will need to instill a feeling that the best strategy available, to each single individual, is to refrain from Exit – productive as well as unproductive – in order to concentrate instead on supplying productive effort and initiative for the socialized sector. Such a change would of course have many implications. Certainly one of the most important is that the private sector is then destined to wither away, much along the lines of the visions for the future of Soviet agriculture that have been quoted on repeated occasions above.

It seems hard indeed to believe that this should be possible to achieve in the near future. Not only does the current wager on the private sector represent the very opposite of reducing the plots to household gardens of the kind normally found in the West. Even the policies of promoting gardening associations and *podkhozy* seem destined to reinforce precisely those mechanisms of *passive* Loyalty which have been seen above to stand at loggerheads with the requirements of an *active* Loyalty. In order to substantiate these claims we shall proceed in two steps. We shall start with the function of the controlling bureaucracy and then proceed to the controlled peasantry and to their urban colleagues.

The Controlling Bureaucracy

In a footnote to his original work on EVL, Hirschman quotes John Hicks as saying that the 'best of all monopoly profits is a quiet life.'[27] The underlying rationale for this quotation would appear to be intimately linked with the seminal idea of EVL. As we may recall, the background story was the apparent riddle of Nigerian railroads being able to continue a grossly inefficient operation in spite of competition from private trucks and buses. The proposed solution was that the availability of an Exit had served to defuse potentially effective Voice against the malfunctions. Consequently, the responsible bureaucrats were given the opportunity to continue their 'soft' existence without being overly bothered by protests and complaints.

Similar conclusions can be reached in our case. Once 'support' of the kind outlined above has been introduced, this also implies that an Exit has been made available. Food can be bought at the markets (by some), a private refuge is available on the plots (for some), and whatever Voice does emanate against the system (from the remainder) can be tolerated by members of the *nomenklatura* who are otherwise busy enjoying a Hicksian 'quiet life'. In all fairness, we should of course note that this reasoning does not imply that bureaucrats do nothing but ride limousines and eat caviar. No doubt some of them work quite hard. The problem, however, concerns the *object* rather than the *volume* of their efforts.

In order to elaborate on this latter point, we may draw a parallel here with Kornai's previously mentioned 'soft budget constraint'. At the focus of his interest lies a distinction between firms that are 'demand-constrained' and those that are 'resource-constrained'. In the former case, which tends to be prevalent in the West, firms cannot expand beyond the limit that is set by demand for their products. If they do, bankruptcy will eventually follow. In the latter case, however, which is typical of the Eastern bloc, demand for their output tends not to be a binding constraint and firms continue to expand as long as they in some way or other can procure the material resources necessary to do so.[28]

The really crucial point is that enterprise managers in the latter case will be less concerned with the actual process of production and its outcome than with various related activities that form the real determinants of the rules according to which rewards and penalties are issued. This would appear to be directly applicable to the activities of bureaucrats controlling agricultural activities. Vasilev has captured it in the following way: 'When faced with a choice between the difficulties of organizational work and the simplicity of a paper solution, our ministerial comrades have decided in favor of the latter.'[29]

This preference for 'paper solutions' is not of course a trait that is specific to agriculture. Its particular relevance in our case, however, is that we are dealing here with the production of food, an area where the reaction of the consumer to any deterioration will be most rapidly felt. The repeated Polish experiences of food price increases producing riots drives this message home rather forcefully. In the absence of those soft productive Exits that have been discussed above, the impact of inefficient management on the system as a whole might have produced dire consequences indeed.

It is of course a matter of taste whether these observations justify giving the bureaucracy the label of a 'quiet life'. We may safely assume, however, that having to learn a completely different *modus operandi* – as implied by the current attempts at *perestroika* – is something that will not be warmly welcomed. The real seriousness of the process derives from the fact that for various reasons it tends to be addictive. An article in *Selskaya Zhizn* has phrased it in the following way: 'Some have grown so used to fixed wages, to guaranteed investment funds, and to free credits, that *khozraschet* to them seems of little meaning.'[30] People who have spent their entire professional life doing things a certain way – the 'paper' way as it were – will have strong vested interests in maintaining precisely that system, simply in order not to devalue their own human capital.

Yanov's account of how the reform attempts in the 1960s offered the rural managerial elite, i.e. *kolkhoz* chairmen and *sovkhoz* directors, a chance to prove conclusively that they functioned better without the controlling Party bureaucracy would also seem to provide additional writing on the wall. At that time, the outcome of the contest, which Yanov claims that Khrushchev consciously allowed to crupt, was that the local Party organ in rural areas, the *raikom*, was simply abolished.[31] The Party *apparatchiki* of today are undoubtedly wary of seeing that story repeat itself. Hence they can be expected to welcome any soft productive Exits that will allow them to continue with business as usual, without having to pay the full price for doing so.

The role of the economic managers, i.e. the farm chairmen and directors, is of course somewhat less clear-cut than that of the controlling bureaucrats. Sometimes they will be in cahoots with local Party officials and may then be assumed to display a fairly similar pattern of behaviour. Other times they will be in opposition to 'petty tutelage' and may then perhaps welcome *perestroika* as a liberator. Depending on the actual situation at hand, we may typically expect both to hold true to some extent. One would probably be hard put to find local managers who are (honestly) all in favour of the new policy.

This, however, is an issue that is somewhat peripheral to our main argument and we shall not pursue it further. Instead, we shall turn now to examine what effects the current wager on the plots may have on the long-run moral and psychological state of individual producers (and 'vacationers'). These latter effects may well be considerably more serious than merely allowing the bureaucracy a 'quiet life', however defined. We have seen above how Stalin's policy of collectivization may actually have led to the emergence of a class of backward conservative *muzhiki*, and it is perhaps not overly speculative to assume that the current wager on the plots may serve to spread and reinforce precisely such sentiments. We shall approach this issue by investigating first the case of the proper peasants and then that of their urban counterparts.

Incentives and Collective Contracts

Much of what has been said above regarding the various soft Exits has been concerned with distortions in or a total lack of incentives for the peasants to work for the collective. Against that background, it is perhaps not surprising to find that a substantial share of labour requirements in the socialized sector of agriculture derives from a perceived need to set up a veritable army of supervisors. In his account of the reform attempts of the 1960s, for example, Yanov claims that the share of supervisors in some *kolkhozy* was as high as 40 per cent of all workers.[32] This certainly gives an indication as to where some of the gains from introducing the *zvenya* were to be found. It is also significant that, if anything, this problem appears to have grown worse over time.

In his speech at Murmansk, in October 1987, Gorbachev claimed that the overall Soviet bureaucracy – which makes up the much criticized 'brake mechanism' – comprises today a total of about 18 million people.[33] According to Shmelev, the size of farm level administration amounts to no less than 1.3 million people.[34] To this we may add a further 0.5 million, in the Gosagroprom and the RAPO hierarchies.[35] These latter millions have been put in the place of the *kulaks*, those hard-working and resourceful peasants who were once so successfully uprooted by Stalin. Their mission is to substitute supervision and control for incentive and initiative. At his May 1988 meeting with leading representatives of the Soviet mass media, Gorbachev addressed precisely this latter problem. In a rather stunning manifestation of *glasnost*, he stated that an 'enormous pyramid burdens the peasant, in whom we still do not have faith.' Speaking further about the need for new forms of organization, he hinted that many 'specialists' might become redundant, since the peasants 'know that these people are parasites'.[36]

Six score and seven years after the emancipation of the Russian serfs, the leader of one of the world's two superpowers admits that the regime still has no confidence in its peasantry, and that the peasants continue to be burdened by a feudal-like pyramid of controlling bureaucrats! If we interpret activating Voice as inducing single individuals to undertake actions in order to support and improve Soviet agriculture, then a large section of this stratum of supervisors will quite obviously have to go. Hence it is rather logical that we have seen provisions recently made for industries to take over and run backward farms. It is also interesting to hear people like Shmelev talk about actually liquidating backward *kolkhozy* and *sovkhozy*.

It is certainly tempting – in this perspective – to see the current promotion of collective and family contracts, linking rewards directly to effort, as an attempt to elicit increased productive initiative without really going all the way to a full dismantling of the socialized sector. We may, however, seriously question the prospects for success with such half-way measures. No matter what emphasis is placed on the allegedly socialist nature of these essentially private undertakings, on the 'organic combination of personal, collective, and common national interests', to quote Shmelev,[37] these experiments will still be taking place within that very same organizational structure which was once purposely devised to hamper and contain peasant initiative.

Various contractual arrangements may certainly alter the formal nature of incentives, and perhaps even provide a token appearance of democracy and self-determination, but it still remains questionable to what extent it will be possible to persuade the human beings inside that system – peasants as well as managers – that something essentially new is actually taking place. Indeed, if we look at the history of past attempts at introducing similar arrangements, it may be difficult to convince anyone that something essentially new is going on.

If we understand the current novelty to lie in the principle of pay by results, then it goes back to a general wage reform in 1961–62, which introduced a system of pay based on piece-work and premia. This system, which is still known as the *akkordno-premialnaya sistema*, was based on payment per unit of output, assessed in relation to work norms and production expenditure.[38] If, on the other hand, we should take the novelty to lie on the organizational side, in allowing small groups of people to work under self-determination, then we can trace it back to the late 1940s, in the form of the *beznaryadnye zvenya* which are claimed by Yanov to have played such an important role in the reform movement during the Khrushchev years.[39]

Ideally, we would of course like to see both autonomy and pay by results, but what is happening today resembles neither of these. Indeed, as Shmelev has pointed out there is currently a considerable terminological confusion regarding the different forms of labour contracting, a confusion which may perhaps be interpreted as a sign that it is chiefly the labels and not the actual content that is changing.[40] It is quite evident from a number of sources, however, that there is a distinct willingness to allow considerable scope for new forms of labour organization and remuneration. At the focus of attention is a system which is known as *kollektivnyi podryad*, or 'collective contracts'. Although its first high-level endorsement was given by Brezhnev, at the time of the launching of the 1982 Food Programme,[41] the real take-off for this system can be traced to a March 1983 meeting of the Politburo. Here Gorbachev, who was then Central Committee Secretary in charge of agriculture, announced that collective contracts should be widely introduced during the 1981–85 five-year plan.[42] Quite predictably, a considerable spurt of activity followed.

If we look simply at official Soviet figures, it is no doubt the case that progress with the various forms of collective contracts has been rapid indeed in recent years. Following the impetus of the 1983 endorsement, close to half the field brigades and more than a third of the livestock teams had already been transferred to the new system by the following year.[43] At the beginning of 1987, Central Committee Secretary Viktor Nikonov indicated that some 11 million farm workers, 75 per cent of the ploughed land, and 60 per cent of socialized sector livestock were under collective contracts.[44] Finally, in a September 1987 decree it was stated that a full transfer to collective contracts should be completed by the end of 1988.[45]

As usual, however, there are good reasons for suspecting that there may be a rather poor correspondence between reality and the official façade. Shmelev, for example, frankly notes that frequently 'collective contracts have been introduced not in reality, but only on paper',[46] while Roy and Betty Laird focus on the large number of terminated or abandoned contracts and conclude that 'with the apparent flagrant violation of the agreements, there is reason to believe that many of the contracts that survive exist largely on paper.'[47] Such views also find support at the very top. In a speech to the Central Committee in January 1987, for example, Secretary Nikonov voiced considerable scepticism: 'When looking at the figures (those nominally working under collective contracts), it would appear that the march of labor contracting ... is victorious; however, the high and highest productivity of labor on the proposed scale is not there.'[48]

In broad terms, the issue at stake here can be interpreted as one of freedom and initiative by decree. There is certainly plenty of evidence on how 'normless' experiments have shown positive results in the past, but then again who has ever heard of a Soviet experiment that failed? It is one thing to allow a small group of strong and motivated *kolkhozniki*, under the auspices of a benign *kolkhoz* chairman and a tolerant *raikom* secretary, to engage in contract work, but it is certainly a different matter altogether to decide that by a certain given date *everyone* should do so.

The very minute that the new scheme is turned into a campaign will also be the moment when we may expect all those activities to materialize which normally accompany Soviet campaigns. What the outcome of that might be has been captured in a recent article by V. P. Gagnon:

> To someone whose whole career has been spent trying to fulfil piece-work related norms, the idea of pay linked to nothing but final output may be incomprehensible. And leaving the implementation of the plan to voluntary actions by the farm workers in the team may seem like folly if the manager has an incomplete understanding of the reform.[49]

On the one hand, we are dealing here with farm managers who are loathe to take the risk of introducing new forms of labour organization in a situation where the potential rewards are at best uncertain. For very much the same reasons that prompt local Party officials to excercise their 'petty tutelage' over the managers, the outcome of the latter's preference for playing safe will be continuous downward interference and violations of signed contracts. Consequently, we have no problem agreeing with Wädekin's conclusion regarding the 1983–84 wave of contracts: 'all in all, the autonomy of the subunit remains a fiction'.[50] This would still seem to be true. We have here the essence of the 'quiet life' that was ascribed above to the controlling bureaucracy. Their activities may certainly at times be quite hectic, but systematically they will be based on the very same *modus operandi*, on the 'paper solution' that was indicated by Vasilev. As a result there follows the need for such support that has been discussed above.

Similarly, we may predict that peasants will be loathe to enter into contracts which in essence boil down to foregoing the guaranteed money wage in return for some very vague possibilities of making money in relation to skills and effort expended. We have here not only the problem of uncertainty regarding inputs. There is also increasing evidence that contracts may at any time be broken by management. Consequently, it is hardly surprising to see Wädekin find that 'many members of brigades and teams show more interest in the payments made for work in the course of the year, which are not tied to final output'.[51] From a narrow economic point of view, simple risk aversion will of course explain much of the poor results of the 'march of labour contracting'. However, it is only by taking into account the underlying social and psychological explanations that we will begin to understand the real nature of this risk aversion.

What we understand as an activation of Voice in effect implies that a once and for all decision is made on whether one wishes to re-create a private peasantry, if such a thing ever existed in Russia, or to promote the formation of true farming cooperatives, in the sense of, say, the Israeli *kibbutzim*. The former case would in practice imply not only a re-admittance of the hated *kulaks*, but also a need to allow the formation of truly autonomous marketing and purchasing cooperatives, in their Western interpretation rather than along the lines of the new Soviet law on cooperatives.[52] It is hard indeed to see this happening, and not just because of opposition from the controlling Party *apparatchiki*. In a recent interview, for example, Tatyana Zaslavskaya stated the following, in response to a question about the possibilities of breaking up the *kolkhozy*:

> I do not know ... of any people who are interested in this. I have been to many *kolkhozy* and talked to many people and those who would give me the chance, who dream of working individually, they just do not exist.[53]

If, on the other hand, we were to look for a successful revival of agriculture *within* the framework of the existing collective structure, then this highlights the need to produce peasants who are actually interested in working for the common good, rather than joining just for the 'free lunch'. We may perhaps even pose the question as to whether it is actually feasible, in *any* type of social and political setting, to build a national agriculture on cooperative principles alone. Israeli experience, for example, strongly suggests that the key to the success of the *kibbutzim* lies in a highly selective recruitment policy, aimed at finding only those individuals who are possessed with very strong personal sympathies for the collective way of life.[54]

To put it very simply, successful agricultural reform is not only a matter of finding technical solutions to problems in the formal incentive system. It also requires human beings who possess both the skills and the interests that are needed for a successful working of the land. It has been the purpose of the presentation of our mechanism of passive Loyalty to show how seven decades of Bolshevik agricultural policy have produced a serious erosion of both of these essential ingredients.

The real anomaly of private-*cum*-socialist agriculture lies in its peculiar combination of the worst of two possible worlds. Security and free-riding in the socialized sector erode the economic preconditions for truly productive initiative in the private sector, whereas such 'private property instincts' that are bred by activities on the plots erode the social preconditions for successful cooperation on the collective fields. What is needed is nothing less than a restoration of such human capital that once existed – a 'return to the soil' to quote Vasilev. If this is at all possible, it will certainly prove to be a rather lengthy process. As academician Tikhonov has put it, we may be dealing here with a period of about 25 years, corresponding to the change of at least two generations of Soviet leaders.[55]

With this rather sombre outlook for the future of Soviet socialized agriculture, let us proceed to investigate what help can be expected from the urban 'recruits'.

The Urban Peasants

With the recent rehabilitation of Chayanov, and mention of Soviet editions of his works,[56] it might be of interest to recall here that in addition to his main contribution on the theory of the family farm, he also wrote in 1920, under the pseudonym of Ivan Kremnev, a charming little piece called the *Journey of My Brother Alexei to the Land of Peasant Utopia.*[57] Written in the midst of War Communism, the story had an obvious political message and quickly became a biblio- graphical rarity. The main plot is the awakening of Alexei Kremnev in Moscow of 1984, a world entirely different from that we know today. In 1934, the Bolsheviks had lost power to the Peasant Labor Party, and a massive decentralization had resulted. Moscow had been reduced to no more than 100,000 inhabitants and other remaining urban centres did not exceed 10,000. Factories had been relocated into the countryside, where small family plots, organized in cooperatives, were lined up back to back to cover the entire expanse of land between the 'cities'.

With the recent rapid growth of the private fringe – Nikitin even speaks of an 'unprecedented dacha boom' around the major cities[58] – this utopia suddenly acquires a rather peculiar relevance. If taken seriously, the analogy is obviously a far-fetched one but it does bring out an issue which may well be of some considerable importance. If we take into account the extensive peasant heritage of the urban population it is perhaps not unrealistic to assume that the current growth of the private fringe will work to produce a disconnection of part of the urban population, from the official sphere of Soviet society, which is similar in principle if not in magnitude to that of the proper peasantry. As was the case with Kremnev's peasants, would it perhaps not be reasonable to assume that the beliefs and values of those active on the fringe will be different from the message of *agitprop* and official Soviet ideology?

Irrespective of what may happen in this respect, however, we will still be left with the very tangible disconnection of the hard core of the proper peasants and with the question of how to achieve a 'reconnection'. The term 'activating Voice' has been used here to imply a need for nothing less than a total suppression of all those soft options that have been outlined above. As we may recall, it was precisely these options that served to produce the strategy of Stay-Silence. Consequently, if we are to achieve Hirschman's original process of delaying Exit in order to promote Voice, the soft options must be discarded. This calls for an end to private plot and market activities, with trust and reliance consequently being placed on the socialized sector. It calls for an end to the improductive activities of drinking, apathy and cynicism, with effort instead being supplied for the good of the collective. Above all, however, it calls for Voice of the constructive kind to be forthcoming within channels that belong to and may be monitored by the official sphere of Soviet society.

Needless to say, these requirements are not only contrary to everything that has been said above. They also harbour a rather ominous threat to the Soviet regime. If such a transformation were actually to occur, this would mean jettisoning all those mechanisms that in the past have provided Soviet society with its basic political and social stability. What is there to ensure that the new 'model' would not instead provide Voice of the original Hirschman kind, i.e. an intolerable content, expressed in an intolerable form?

In Lieu of a Conclusion

The various conclusions that have emerged from the different stages of our analysis perhaps remove the need to end the presentation with a grand conclusion. Instead, we shall Exit by way of returning to our point of entry – the Pyrrhic victory. If Gorbachev continues on the path of his predecessors, i.e. to accept the price that has to be paid for maintaining those mechanisms of a passive Loyalty which serve to postpone the needs for eventual reform, then he may well be destined for a fate similar to that which once befell Pyrrhus, the dethroned king of Epirus, in his struggle against the Romans.

He may win the battle over the plots, in the sense of eliciting increased output from that quarter. He may win the battle over the gardening associations and over the *dachniki*, in the sense of mobilizing the urban population for the same purpose, and he may even win the battle over the *podkhozy*, in the sense of making industry – and the armed forces – assume a greater responsibility for feeding their own employees. In gaining all of these victories, however, he may well end up losing the war over Soviet socialist agriculture.

All of those 'support' activities which have been the topic of this book are certainly beneficial in the narrow sense of defusing a source of potential unrest, thus allowing the 'Party affair' to continue, and of propping up backward *kolkhozy* and *sovkhozy*, thus allowing the system to maintain an outward Potemkin façade of being socialist and cooperative. In so doing, however, they continuously work to erode the very foundations of the large-scale socialized sector, thus reducing the hope of ever finding a real solution to the 'permanent crisis' to a very low order of probability indeed. As Emile Zola is reported once to have said, '*C'est dur, l'agriculture.*'

Notes

Introduction

1. Jasny, N. (1951) '*Kolkhozy*, the Achilles heel of the Soviet regime', *Soviet Studies* 3 (no. 2).

Preface

1. Hedlund, S. (1984) *Crisis in Soviet Agriculture*, London and Sydney: Croom Helm, New York: St Martin's.
2. *Pravda*, 26 February 1986.
3. *Pravda*, 25 Sept. 1987.
4. Shmelev, N. (1988) 'Novye trevogi', *Novyi Mir*, no. 4, p. 164.
5. *Literaturnaya Gazeta*, 11 May 1988, p. 3.
6. *Pravda*, 12 July 1985.
7. FBIS Daily Report: Soviet Union, 30 January 1987, p. R27.
8. *Pravda*, 25 Sept. 1987.
9. *Moscow News*, suppl., no.4, 1988.
10. Shmelev, N. (1987) 'Avansy i dolgi', *Novyi Mir*, no.6, p. 146.
11. Zalygin, S. (1987) 'Povorot', *Novyi Mir*, no.1, p. 234.
12. *Literaturnaya Gazeta*, 11 May 1985, p. 1.
13. Wädekin, K.-E. (1967) *Privatproduzenten in der sowjetischen Landwirtschaft*, Köln: Verlag Wissenschaft und Politik. With the exception of a number of appendices which appear only in the original German version, references will be made to a revised English edition in Wädekin, K.-E. (1973) *The Private Sector in Soviet Agriculture*, Berkeley, CA: University of California Press.

Chapter 1. A black hole in the Soviet economy

1. Wädekin, K.-E. (1973) *The Private Sector in Soviet Agriculture*, Berkeley, CA: University of California Press, p. xv.
2. Laird, R.D. and Crowley, E. (eds) (1965) *Soviet Agriculture: The Permanent Crisis*, New York: Praeger.

3. This section relies heavily on Hedlund, S. (1984) *Crisis in Soviet Agriculture*, London and Sydney: Croom Helm, New York: St Martin's.
4. Marx, K. and Engels, F. (1977) *Manifesto of the Communist Party*, Moscow: Progress, p. 4.
5. This is argued in greater detail in Hedlund (1984), Chap. 2.
6. Lewin, M. (1974) *Political Undercurrents in Soviet Economic Debates: From Bukharin to the Modern Reformers*, Princeton: Princeton University Press, p. 318.
7. Yanov, A. (1984) *The Drama of the Soviet 1960s. A Lost Reform*, Berkeley, CA: Institute of International Studies, p. 22.
8. See further Severin, B.S. (1987) 'Solving the Soviet livestock feed dilemma: key to meeting food program targets', in: US Congress (1987) Joint Economic Committee, *Gorbachev's Economic Plans*, Vol. II, Washington, DC: Government Printing Office, pp. 47–48, and *passim*. A considerable improvement has taken place on this account in recent years, but feed conversion ratios still remain very poor, exceeding those of the United States, for example, by some 40 per cent (USDA (1986) *USSR: Situation and Outlook Report*, Washington, DC: Government Printing Office, pp. 15–16).
9. For the period 1976–80, see Hedlund (1984), p. 88. Plan targets for 1981–85 (238– 43 million tons) and for 1986–90 (250– 55 million tons) can be found in *Pravda*, 25 May 1982, while the actual results for 1981–85 are given in Narkhoz (1986) *Narodnoe khozyaistvo SSSR v 1985* g., Moscow: Finansy i statistika, p. 180. Harvest figures for 1986–87 are still somewhat obscure. *Pravda*, 19 November 1987, claimed that both years exceeded 210 million tons and that the increase over 1981–85 average was 17 per cent. With a reported figure of 210.1 million tons for 1986, this would imply about 212 million tons for 1987. See further *RSEEA Newsletter*, vol. 9, no. 4, p. 1.
10. Narkhoz (1963) *Narodnoe khozyaistvo v SSSR v 1962 g.*, Moscow: Finansy i statistika, p. 6. Narkhoz (1987) *Narodnoe khozyaistvo SSSR za 70 let: Yubileinyi statisticheskii ezhegodnik*, Moscow: Finansy i statistika, p. 5.
11. Population data in Narkhoz (1966) *Narodnoe khozyaistvo SSSR v 1965 g.*, Moscow: Finansy i statistika, p. 7; Narkhoz (1976) *Narodnoe khozyaistvo SSSR v 1975 g.*, Moscow: Finansy i statistika, p. 7; Narkhoz (1981) *Narodnoe khozyaistvo SSSR v 1980g.*, Moscow: Finansy i statistika, p. 7; Narkhoz (1986), p. 5.
12. See further McCauley, M. (1976) *Khrushchev and the Development of Soviet Agriculture: The Virgin Lands Programme, 1953–64*. London: Macmillan, Chap. 4 and *passim*.
13. Ibid., p. 82. Much of this addition, however, was quickly lost. The 'true' total came to no more than 32.7 million hectares.
14. Narkhoz (1968) *Narodnoe Khozyaistvo SSSR v 1967 g.*, Moscow: Finansy i statistika, p. 348, Narkhoz (1987), p. 222.
15. Narkhoz (1987), p. 275. The definition used here is *po vsemu*

kompleksu rabot, which includes investment in research, water management and material supply organizations, but excludes the processing and machine-building industries. A narrow definition of the farm sector only produces a figure around 20 per cent, while the entire 'agro-industrial complex' accounts for more than a third of total investment.

16. See further Hedlund (1984), Chap 6.
17. In the case of *sovkhozy*, a similar development can be observed, with the value of output produced per 1,000 rubles of fixed assets falling from 1,294 rubles in 1970 to merely 650 rubles in 1980 (Narkhoz (1987), p. 292).
18. Doolittle, P. and Hughes, M. (1987) 'Gorbachev's agricultural policy: Building on the Brezhnev food program', in: US Congress (1987), p. 28.
19. Goodman, A., Hughes, M. and Schroeder, G. (1987) 'Raising the efficiency of Soviet farm labor: Problems and prospects', in: US Congress, pp. 108–9.
20. Yanov (1984), p. 9.
21. Goodman, Hughes and Schroeder (1987), p. 102.
22. Johnson, D.G. and Brooks, K. McConnell (1983) *Prospects for Soviet Agriculture in the 1980s*, Bloomington, IN: Indiana University Press, p. 166.
23. Narkhoz (1986), p. 304.
24. Manevich, E. (1981) 'Rationalsnoe ispolzovanie rabochei sily', *Voprosy Ekonomiki*, no. 9, p. 60.
25. Yanov (1984), pp. 116–20.
26. Gorbachev on Moscow television (FBIS Daily Report: Soviet Union, 10 November 1987).
27. Prodovolstvennaya (1982) *Prodovolstvennaya programma SSSR na period do 1990 goda i mery po ee realizatsii. Materialy maiskogo plenuma TsK KPSS*, Moscow: Politizdat.
28. At the time, its magnitude was kept a secret, and during the years 1981–85, no data on Soviet grain harvests were made official. Much to the surprise of Western observers, however, and no doubt partly as a consequence of the relatively favourable 1986 crop, the Soviet leadership recently decided to come clean, releasing the previously secret data. Thus, according to the 1986 edition of the statistical annuary *Narodnoe khozyaistvo*, the 1981 grain harvest reached no more than 158.2 million tons (Narkhoz (1986) p. 180). The target was 220.0 million tons.
29. Malish, A. (1982) 'The food program: A new policy or more rhetoric?' in: US Congress (1982) Joint Economic Committee, *Soviet Economy in the 1980s: Problems and Prospects*, vol. II, Washington, DC: Government Printing Office.
30. Narkhoz (1986), p. 180. Prodovolstvennaya (1982), p. 33.
31. Wädekin, K.-E. (1987) 'Agriculture', in M. McCauley (ed.), *The Soviet Union under Gorbachev*, New York: St Martin's, p. 122.
32. See e.g. Gray, K. R. (1987) 'Reform and resource allocation in

Soviet agriculture', in: US Congress, p. 10, footnote 2 and Wädekin, K.-E. (1987) 'Commentary', in US Congress, pp. 131–2.
33. White, S. (1986) 'Economic performance and Communist legitimacy', *World Politics* 3, no. 3, p. 481.
34. Prodovolstvennaya (1982), pp. 18–19.
35. Suslov, I. (1982) 'Kolkhozy v sisteme narodnogo khozyaistva', *Voprosy Ekonomiki*, no. 12, p. 27.
36. Kornai, J. (1979) 'Resource-constrained versus demand-constrained systems', *Econometrica* 47, no. 4.

Chapter 2. The emergence of a 'support' agriculture

1. In a report presented at the time of the introduction of the 1982 Food Programme, then Premier Tikhonov indicated that as much as 20 per cent of total agricultural output was lost every year in the process of procurement, transport and storage (Tikhonov, I. (1982) 'Edinyi narodnokhozyaistvennyi kompleks mnogonatsionalnogo sovetskogo gosudarstva', *Kommunist*, no. 11, p. 4). This problem will be returned to at greater length later.
2. Conquest, R. (1986) *The Harvest of Sorrow: Soviet Collectivization and the Terror Famine*, New York and Oxford: Oxford University Press. Lewin, M. (1965) 'The immediate background to Soviet collectivization', *Soviet Studies* 17, no. 2. Lewin, M. (1966) 'Who was the Soviet Kulak', *Soviet Studies* 18, no. 2. Lewin, M. (1968) *Russian Peasants and Soviet Power: A Study of Collectivization*, London: Allen and Unwin. The emergence of the *kolkhoz* system as a whole is described in Hedlund, S. (1984) *Crisis in Soviet Agriculture*, London and Sydney: Croom Helm, New York: St Martin's, Chap. 2.
3. It is significant that the peasants increasingly came to interpret the letters VKP as *Vtoroe Krepostnoe Pravo* (second serfdom), rather than in their proper meaning of *Vsesoyuznaya Kommunisticheskaya Partiya* (All-union Communist Party) (Conquest (1986), p. 152).
4. Raig, I.Kh. (1984) 'Razvitie lichnogo podsobnogo khozyaistva v sovetskom derevne', *Istoryia SSSR*, no. 5, p. 123.
5. See Volin, L. (1970) *A Century of Russian Agriculture*. Cambridge, MA: Harvard University Press, pp. 244–8.
6. Lewin, M. (1985) *The Making of the Soviet System. Essays in the Social History of Interwar Russia*. New York: Pantheon Books, p. 179.
7. Raig (1984), p. 123.
8. Ibid.
9. Shmelev, G.I. (1983) *Lichnoe podsobnoe khozyaistvo: vozmozhnostii i perspektivy*, Moscow: Politicheskaya Literatura, p. 7.
10. Ibid., p. 8.
11. Raig (1984), p. 123. See also Ostrovskii, V.B. (1967) *Kolkhoznoe krestyanstvo SSSR. Politika partii v dervne i ee sotsialno-ekonomischeskie resultaty*, Saratov: Izdatelstvo saratovskogo universiteta, p. 73.

12. It is an interesting reflection of Gorbachev's *glasnost*, however, that the 1987 – jubilee – edition of *Narodnoe khozyaistvo* contains a number of tables with statistics for the pre-war period, statistics which clearly reflect the impact of collectivization (e.g. Narkhoz (1987) *Narodnoe khozyaistvo SSSR za 70 let: Yubileinyi statisticheskii ezhegodnik*. Moscow: Finansy i statistika, pp. 208, 210, 253, 258).
13. Raig (1984), p. 123. These rights were codified at the end of the decade. See further Belyanov, V.A. (1970) *Lichnoe podsobnoe khozyaistvo pri sotsializme*, Moscow: Nauka, pp. 147–53.
14. Shmelev, G.I. (1964) *Raspredelenie ispolzovanie truda v kolkhozakh*. Moscow: Mysl, p. 136.
15. Raig (1984), p. 123.
16. See further Wädekin, K.-E. (1973) *The Private Sector in Soviet Agriculture*, Berkeley, CA: University of California Press, Chap. 8.
17. E.g. de Pauw, J. W. (1969) 'The private sector in Soviet agriculture', *Slavic Review* 28, no. 1, p. 68.
18. Wädekin argues convincingly that Khrushchev's statements were seriously meant, and that he believed them to be realistic: 'Everything points to the fact that he himself at the time really believed that it was possible to reach this target, even though it appeared utopian from the very beginning to sensible and competent observers' (Wädekin (1973), p. 272).
19. Ibid., p. 265.
20. Ibid., pp. 268–9.
21. Ibid., p. 271.
22. Ibid., pp. 275–6.
23. Ibid., pp. 284–5.
24. Ibid., p. 282.
25. Volin (1970), pp. 228–9.
26. Wädekin (1973), p. 282.
27. 'O merakh borby s raskhodovaniem iz gosudarstvennykh fondov khleba i drugikh prodovolstvennykh produktov na korm skotu' (*Pravda*, 28 August 1956).
28. One Soviet source indicates that in 1975 Soviet pigs devoured no less than 1.4 million tons of bread, corresponding to 5–6 kilograms of bread *per capita* of the human population, or to about 4 per cent of the total of bread and grain products sold that year (quoted by Lane, A. (1982) 'USSR: Private Agriculture on Center Stage', in US Congress (1982) Joint Economic Committee, *Soviet Economy in the 1980s: Problems and Prospects*, vol. II, Washington, DC: Government Printing Office, p. 28).
29. *Pravda*, 7 May 1985.
30. Wädekin (1973), pp. 293–4.
31. Ibid., p. 305.
32. Ibid., p. 289.
33. Ibid., pp. 309-10.
34. Ibid., p. 314.

35. This interpretation has been suggested to the author by Karl-Eugen Wädekin.
36. Wädekin (1973), p. 315.
37. Raig (1984), p. 126.
38. The first of these, 'O lichnykh podsobnykh khozyaistvakh kolkhoznikov, rabochykh, sluzhashchikh i drugikh grazhdan i kollektivnom sadovodstve i ogorodnichestve', was issued on 14 September 1977 (for the full text see Leninskaya (1978) *Leninskaya agrarnaya politika KPSS: sbornik vazhneishikh dokumentov (mart 1965 g.–ijul 1978 g.)*, Moscow: Politizdat, pp. 632–9), and the second, 'O dopolnitelnykh merakh po uvelicheniyu proizvodstva selskokhozyaistvennoi produktsii v lichnykh podsobnykh khozyaistvakh grazhdan', on 8 January 1981 (Leninskaya (1983) *Leninskaya agrarnaya politika KPSS: sbornik vazhneishikh dokumentov (avgust 1978 g.–avgust 1982 g.)*, Moscow: Politizdat, pp. 492–505). The texts of both decrees have been reprinted as appendices in Kalinkin, A.F. (1981) *Lichnoe podsobnoe khozyaistvo, kollektivnoe sadovodstvo i ogorodnichestvo*, Moscow: Kolos. Although primarily aimed at providing support for the private sector 'narrowly defined', these first two decrees also had a lot to say of such private orchard and gardening activites that are discussed further in Chapter 3. Yet there would be issued, on 15 May 1986, an additional decree, 'O merakh po dalneishemu razvitiyu kollektivnogo sadovodstva i ogorodnichestva' (*Izvestia*, 7 June 1986), in support of precisely such activities. The latter illustrates the growing importance that is being placed on the private fringe.
39. 'O dopolnitelnykh merakh po razvitiyu lichnykh podsobnykh khozyaistv grazhdan, kollektivnogo sadovodstva i ogorodnichestva' (*Pravda*, 25 September 1987). The decision to review restrictions on private sector activities was taken at a meeting, on 23 July 1987, of the Politburo (*Izvestia*, 26 July 1987). A draft of the text to the *kolkhoz* charter (*Primernyi ustav kolkhoza*) was published in *Ekonomicheskaya Gazeta*, no. 3, 1988, pp. 15–18. The draft text was discussed at a specially convened Fourth All-Union Kolkhoz Congress, held in March. Gorbachev's speech to that congress can be found in *Ekonomicheskaya Gazeta*, no. 13, pp. 1–7. For comments, see also Hanson, P. (1988) 'Gorbachev addresses Kolkhozniks Congress', *Radio Liberty Research*, 131/88, 24 March.
40. Conquest (1986), p. 164.
41. Shmelev (1983), p. 14.
42. Shmelev, G.I. (1984) *Podsobnye khozyaistva predpriyatii i naseleniya*, Moscow: Znanie, p. 24.
43. *Literaturnaya Gazeta*, 12 March 1980, p. 10.
44. Bradley, M.E. (1971) 'Incentives and labor supply on Soviet collective farms', *Canadian Journal of Economics* 4, no. 3, p. 349.
45. Shmelev, G.I. (1981) 'Obshchestvennoe proizvodstvo i lichnoe podsobnoe khozyaistvo', *Voprosy Ekonomiki*, 53, no. 5, p. 69. See also Sidorenko, V.I. (1985) 'Vazhnyi istochnik popolneniya

prodovolstvennogo fonda', *Planirovanie i uchet v selskokhozyaistvennykh predpriyatiyakh*, no. 3.

46. Data on recent development seem to suffer from inconsistencies in the Soviet sources. Wädekin (1985) 'Private gardeners in the USSR', *Radio Liberty Research*, RL 174/85, 30 May, p. 7, discusses these and also presents somewhat different data. The differences, however, are minor only, with e.g. 25.8 per cent being given for 1979.
47. Konstitutsiya (1977) *Konstitutsiya (osnovnoi zakon) soyuza sotsialisticheskikh respublik*. Moscow: Politicheskay a Literatura, p. 10.
48. It may be relevant in this context to mention that in the previous, 1936, Constitution, plots in *kolkhozy* were treated as distinct from those provided to other categories of the population, Article 7 dealing with the former and Article 10 with the latter (Konstitutsiya (1961) *Konstitutsiya (osnovnoi zakon) soyuza sotsialisticheskikh respublik*, Moscow: Politicheskaya Literatura pp. 11–13).
49. Kalinkin, A.F. (1981) *Lichnoe podsobnoe khozyaistvo, kollektivnoe sadovodstvo i ogorodnichestvo*, Moscow: Kolos, p. 11. Total land holdings in the respective categories amounted to 4.3, 2.2 and 1.2 million hectares.
50. Kalinkin, A.F. (1982) 'Razvitie lichnogo podsobnogo khozyaistva', *Ekonomika selskogo khozyaistva*, no. 4, p. 64.
51. Primernyi (1970) *Primernyi ustav kolkhoza*, Moscow: Kolos. An English translation of the text can be found in *Current Digest of the Soviet Press*, no. 50, 1970.
52. These differences are discussed in Wädekin (1973), pp. 344-66.
53. Wädekin, K.-E. (1988) ('The new *kolkhoz* statute: A codification of restructuring on the farm', *Radio Liberty Research*, RL 36/88, 28 Jan.) discusses the most important differences between the 1969 charter and the draft text of the 1988 successor.
54. Another important distinction between the *kolkhozniki* and the rest of the population lies in the fact that it is only in the former case that the size of the plot can (legally) be made dependent on labour participation in the socialized sector (Pravovoi (1984) *Pravovoi rezhim zemel v SSSR*, Moscow: Nauka, p. 311).
55. Primernyi (1970), p. 18.
56. This point is obviously connected to the current discussion about a new scheme – known as *arendnyi podryad* – which allows the peasants to lease land. This potentially important change was heralded by Gorbachev himself at the October 1987 Central Committee plenum (*Ekonomicheskaya Gazeta*, no. 44, p. 2).
57. Articles 42–4.
58. See further Kozyr (1981) 'Pravovye osnovy vedeniya lichnogo podsobnogo khozyaistva', in Kalinkin, pp. 44–8, or Wädekin (1973), pp. 31–8. The legal basis for making allotments of land to such people, who are not ideologically considered as 'peasants', can be found in Article 26 of the Basic Land Law of the USSR (*Osnovy zemelnogo zakonodatelstva soyuza SSR i soyuznykh respublik*). See

e.g. Novoe (1969) *Novoe v zemelnom zakonodatelstve*, Moscow: Yuridicheskaya Literatura, pp. 65–6.

59. These figures are for the RSFSR. Pravovoi (1984), p. 313, recognizes five distinct categories, the fifth being composed of generals, admirals and senior officers. The former two may receive plots up to 0.25 hectares, while the latter are entitled to no more than 0.15 hectares.
60. Kozyr (1981), p. 72.
61. For the sake of completeness we ought perhaps to mention as well the special category of 'service strips' (*sluzhebnyi nadel*), small plots of land that are placed at the disposal of people living and working far from populated areas, such as loggers and certain groups of construction and railroad workers. While the simple existence of this category surely adds to our overall perspective of the 'support' function of private agriculture, it is of minute importance and shall not figure in our discussion below. See further Kalinkin (1981), pp. 48–50, and Pravovoi (1984), pp. 294–8.
62. Shmelev (1983), p. 6.
63. Lewin (1985), p. 180.
64. Narkhoz (1987), p. 222. The total figure for the private sector, i.e. including land under buildings, roads, etc., stands at 8.55 million hectares. In 1985, 1.3 million hectares of that area was devoted to fruit and berries, corresponding to about 41 per cent of the total for the nation as a whole (Narkhoz, 1986, p. 217).
65. Shmelev (1983), p. 11; Nove, A. (1977) *The Soviet Economic System*, London: Allen and Unwin, p. 123. Other Soviet sources seem to favour the higher figure (e.g. Kalinkin, 1982, p. 67).
66. Schinke, E. (1972) 'Soviet agricultural statistics', in Treml, V. and Hardt, J. (eds), *Soviet Economic Statistics*, Durham, NC: Duke University Press, p. 240.
67. Lane, A. (1982) 'USSR: Private Agriculture on Center Stage', in US Congress (1982), p. 28. Unfortunately no year is given for this estimate, so it is compared to official Soviet figures for 1980.
68. Shmelev (1983), p. 25.
69. Shmelev (1985) 'Lichnoe podsobnoe khozyaistvo', *Novoe v zhizni nauke i tekhnike, Seriya Ekonomika*, no. 3, p. 3.
70. Novye (1987) 'Novye yavleniya v lichnom podsobnom khozyaistve', *Voprosy Ekonomiki*, no.7, p. 139.
71. Failure by the entire household to participate in work in the socialized sector is one of the causes that may lead to a loss of plot rights.
72. Kalinkin (1981), p. 13.
73. Shmelev (1981), p. 68.

Chapter 3. The two fringes of Soviet agriculture

1. In the wake of the 1982 Food Programme, *Krasnaya Zvezda*, the central daily of the Ministry of Defence, had a wide coverage of

how the armed forces made their contribution towards solving the country's food problem.

2. In earlier times, the *podkhozy* were referred to as *orsy*, an acronym for *otdel rabochego snabzheniya*, or department for supply to workers, which reflects that their primary function originally was that of providing food for urban workers during times of emergency (Volin, L. (1970) *A Century of Russian Agriculture*, Cambridge, MA: Harvard University Press, p. 535).
3. See further Shmelev, G.I. (1984) *Podsobnye khozyaistva predpryiatii i naseleniya*. Moscow, Znanie, pp. 7–8.
4. Quoted in Volin (1970), p. 289.
5. Ibid.
6. *Izvestia*, 23 Dec. 1978. See also Shmelev (1982), p. 8.
7. Prodovolstvennaya (1982) *Prodovolstvennaya programma SSSR na period do 1990 goda i mery po ee realizatsii. Materialy maiskogo plenuma TsK KPSS*, Moscow: Politizdat, p. 43.
8. Ibid., p. 16.
9. Ioffe, M. (1982) 'Podsobnye khozyaistva promyshlennykh predpriyatii i organisatsii', *Planovoe khozyaistvo*, no. 11, p. 73.
10. Ibid., p. 74. *Pravda*, 30 May 1980 gives a number of examples of the performance of *podkhozy* in various regions.
11. Shmelev (1984), pp. 13–14.
12. It is symptomatic, for example, that when a model charter for the *podkhoz* was published in 1986 it was explicitly stated, in the very first clause, that it did not apply to the military *podkhozy*, nor to those belonging to the KGB or the Ministry of the Interior (Primernoe (1986) 'Primernoe polozhenie o podsobnom selskom khozyaistve, yavlyayushchemsya strukturnym podrazdeleniem predpriyatiya, organizatsii, uchrezhdeniya', *Ekonomika selskogo khozyaistva*, no. 5, p. 88).
13. *Krasnaya Zvezda*, 15 June 1982.
14. *Krasnaya Zvezda*, 11 July 1982.
15. *Krasnaya Zvezda*, 13 June 1982.
16. *Krasnaya Zvezda*, 6 June 1982.
17. Shmelev (1984), p. 15.
18. *Pravda*, 9 June 1980.
19. *Trud*, 25 January 1981.
20. *Pravda*, 9 June 1980.
21. Ioffe (1982), p. 72.
22. *Izvestia*, 25 December 1981.
23. *Izvestia*, 26 December 1981.
24. *Pravda*, 25 September 1987.
25. Tenson, A. (1983) 'Factory farms or farm factories', *Radio Liberty Research*, RL 71/83, 8 February, p. 4
26. Ioffe (1982), pp. 75–6.
27. Shmelev (1984), p. 16.
28. Ibid., pp. 10–12.
29. Narkhoz (1986) *Narodnoe khozyaistvo SSSR v 1985 g.*. Moscow:

Finansy i statistika, pp. 277, 285.
30. *Trud*, 25 January 1981.
31. Narkhoz (1987) *Narodnoe khozyaistvo SSSR za 70 let: Yubileinyi statisticheskii ezhegodnik*, Moscow: Finansy i statistika, p. 299.
32. Ibid., pp. 253, 299.
33. Shmelev (1984), p. 16.
34. Ioffe (1982), p. 75.
35. Narkhoz (1987), pp. 258, 299.
36. The main characteristics of these two forms of gardening are described in (1973), Selskokhozyaistvennaya (1972; 1973) *Selskokhozyaistvennaya entsiklopediya*, Moscow: Sovetskaya eutsiklopediya, (1972), pp. 96–7; (1973), pp. 395–6.
37. Kubyak, L.A. and Maksakova, G.P. (1981) 'Kollektivnoe sadovodstvo ogorodnichestvo i zhivotnovodstvo rabochikh i sluzhashchikh', in: Kalinkin, A.F. (ed.), *Lichnoe podsobnoe khozyaistvo, kollektivnoe sadovodstvo i ogorodnichestvo*, Moscow: Kolos, p. 139.
38. Ibid., p. 152.
39. See Pravovoi (1984) *Pravovoi rezhim zemel v SSSR*, Moscow: Nauka, p. 101.
40. Ibid.
41. Ibid., p. 108, footnote 5.
42. Ibid., p. 102.
43. Selskokhozyaistvennaya (1972), p. 96; (1973), p. 395.
44. Pravovoi (1984), p. 103.
45. Some of the distinctions made above are briefly summarized in Shmelev, G.I. (1985) 'Lichnoe podsobnoe khozyaistvo', *Novoe v zhizni, nauke i tekhnike. Seriya Ekonomika*, no. 3, p. 27, whereas a more exhaustive account can be found in Kubyak and Maksakova (1981). The common gardens and the cattle associations are discussed in ibid., pp. 152–7. The legal aspects of the cattle associations can be found in Pravovoi (1984), pp. 105–8.
46. Raig, I.Kh. (1986) 'Chto mozhet individualnoe khozyaistvo?' *Sotsiologicheskie Issledovaniya*, no. 1, p. 37.
47. *Izvestia*, 23 Jan. 1982.
48. Shmelev (1984), p. 35.
49. Shmelev (1985), p. 28.
50. Kubyak and Maksakova (1981), p. 130.
51. *Izvestia*, 23 January 1982.
52. Shmelev (1983), p. 25.
53. Shmelev (1985), p. 3.
54. Karakhanova, T.M. and Patrushev, V.D. (1983) 'Kollektivnoe ogorodnichestvo i sadovodstvo—reserv prodovolstvennogo obespecheniya', *Sotsiologicheskie Issledovaniya*, no. 2, pp. 83–4.
55. Shmelev (1985), p. 28.
56. Raig (1986), p. 35.
57. Narkhoz (1986) *Narodnoe khozyaistvo SSSR v 1985 g*. Moscow: Finansy i statistika, p. 220.

58. SSSR (1987) *SSSR v tsifrakh v 1986 g*. Moscow: Finansy i statistika, p. 131.
59. For the sake of curiosity, this exercise will be ended by noting that there was in 1981, in the RSFSR, a total of 117,000 members of cattle associations and 226,000 members of rabbit associations (Shmelev (1985), p. 31).
60. *Izvestia,* 7 June 1986. Although *Izvestia* mentions only the orchard plots (*sadovodcheskie*), Narkhoz (1986), p. 220, and Narkhoz (1987), p. 237, refer to both categories.
61. See further *Literaturnaya Gazeta,* 20 January 1982, p. 10.
62. Although the Russian term (*ustav*) is the same, there is a significant difference between these and the model *kolkhoz* charter. All farming activities by the non-*kolkhoz* population are regulated in separate republican legislation. The charters mentioned here are derived from such laws.
63. Selskokhozyaistvennaya (1972), p. 96.
64. The legal provisions for making such allotments of land can be found in Article 22 of the Basic Land Law (Novoe (1969) *Novoe v zemelnom zakonodatelstve,* Moscow: Yuridicheskaya Literatura, p. 63). See also Pravovoi (1984), pp. 93–101.
65. Section II, Article 12 of the 1978 model charter (Tipovoi (1981) 'Tipovoi ustav sadovodcheskogo tovarishchestva rabochikh i sluzhashchikh', in Kalinkin, A.F. (ed.), *Lichnoe podsobnoe khozyaistvo, kollektivnoe sadovodstvo i ogorodnichestvo,* Moscow: Kolos, p. 217).
66. *Trud,* 25 August 1978.
67. *Literaturnaya Gazeta,* 22 February 1985.
68. The text of this charter is reprinted in Kutin, E.M. (1974) *Kollektivnoe sadovodstvo i ogorodnichestvo,* Moscow: Yuridicheskaya Literatura.
69. The text of this charter is reprinted as an appendix in Kalinkin (1981), pp. 215–22.
70. See Kubyak and Maksakova (1981), pp. 144–45. Even the size of showers (2.5 m^2) and toilets (1.5 m^2) is controlled. The applicable clause is Section I, Article 8 of the 1978 model charter (Tipovoi (1981), p. 216).
71. Wädekin K.-E.(1973) *The Private Sector in Soviet Agriculture,* Berkeley, CA: University of California Press, p. 39.
72. 'O dopolnitelnykh merakh po razvitiyu lichnykh podsobnykh khozyaistv grazhdan, kollektivnogo sadovodstva i ogorodnichestva' (*Pravda,* 25 September 1987).
73. *Literaturnaya Gazeta,* 9 April 1980, p. 12.
74. The exact interpretation of 'mainly' is discussed further in *Ekonomicheskaya Gazeta,* no. 19, 1987, p. 13.
75. *Trud,* 25 Aug. 1978. On credits for private sector agriculture in general, i.e. both core and fringe, see Ushakov, V.M. (1981) 'Finansirovanie i kreditovanie grazhdan na razvitie lichnykh podsobnykh khozyaistv i zhilizhnoe stroitelstve na sele', in A.F.

Kalinkin (ed.), *Lichnoe podsobnoe khozyaistvo, kollektivnoe sadovodstvo i ogorodnichestvo*, Moscow: Kolos.
76. *Ekonomicheskaya Gazeta*, no. 19, 1987, p. 13.
77. *Izvestia*, 23 January 1982.
78. Shmelev (1984), p. 36.
79. The size of these plots is restricted to between 300–600 m^2, if in urban areas, while in urban type settlements they may reach 700–1,200 m^2 (Pravovoi (1984), p. 320). The legal details regarding the allocation of land to individuals in urban areas can be found in ibid., pp. 317–23.
80. Section II, Article 10 (Tipovoi (1981), p. 216).
81. *Literaturnaya Gazeta*, 3 June 1981, p. 10.
82. The letter was received in reply to a previous article by Nikitin, in *Literaturnaya Gazeta*, 21 November 1979.
83. *Literaturnaya Gazeta*, 3 June 1981, p. 10.
84. Article 44 of the 1969 model *kolkhoz* charter permits the *kolkhoz* to provide plots for workers and 'specialists' (teachers, doctors, etc.) working in rural areas and living on its territory. Only if there is vacant land may such rights be extended to other workers – and to pensioners and invalids – who live on its territory (Primernyi, 1970, pp. 18–19). The latter provision would seem to be the one applicable to the private fringe.
85. *Sovetskaya Rossiya*, 2 August 1987.
86. *Literaturnaya Gazeta*, 23 September 1981, p. 11.
87. *Literaturnaya Gazeta*, 20 January 1982, p. 10.
88. 'Ob ispolzovanii pustuyuzhikh zhilykh domov i priusadebnykh uchastkov nakhodyashchikhsya v selskoi mestnosti', *Sovetskaya Rossiya*, 1 August 1987.
89. *Sovetskaya Rossiya*, 2 August 1987. See also Tenson, A. (1987) 'Easing of restrictions on private plots', *Radio Liberty Research*, RL 71/87, 18 September.

Chapter 4. Soviet attitudes to the private sector

1. Dyachkov, G. and Sorokin, A. (1980) 'Rol lichnogo podsobnogo khozyaistva', *Ekonomika selskogo khozyaistva*, no. 1, p. 69.
2. *Literaturnaya Gazeta*, 23 September 1981, p. 11.
3. Shmelev, G.I. (1983) *Lichnoe podsobnoe khozyaistvo: vozmozhnosti i perspektivy*, Moscow: Politicheskaya Literatura, pp. 21–2.
4. Ibid., p. 22.
5. Ibid.
6. Ibid.
7. Wädekin, K.-E. (1973) *The Private Sector in Soviet Agriculture*, Berkeley, CA: University of California Press, p. 17.
8. Shmelev (1983), p. 24.
9. Khronika (1982) 'Khronika nauchnoi zhizni: Problemy lichnogo podsobnogo khozyaistva', *Izvestiya Akademii Nauk SSSR. Seriya ekonomicheskaya*, no. 3, p. 124.

10. Dyachkov and Sorokin (1980), pp. 63–5.
11. Rutkevich, M.N. and Filippov, F.R.(1970) *Sotsialnye peremeshcheniya*, Moscow, p. 59.
12. Ibid., p. 58.
13. Larionova, K. (ed.) (1963) *Politicheskaya ekonomiya. Uchebnoe posobie*, Moscow, p. 341.
14. Blyakhman, L.S. and Shkaratan, O.I. (1973) *NTR, rabochii klass' intelligentsiya*, Moscow, Politicheskaya Literatura, pp. 209–10.
15. Rutkevich and Filippov (1970), pp. 74–5.
16. Sergeev, S.S. (1956) *Voprosy ekonomiko-statistichesko analiza kolkhoznogo proizvodstva*, Moscow, Selkhozgiz, p. 93.
17. Kolganov, M.V. (1953) *Sobstvennost v sotsialisticheskom obshchestvė*, Moscow: AN SSSR, p. 367.
18. Stalin, J.V. (1952) *Ekonomicheskie problemy sotsializma v SSSR*, Moscow: Gospolitizdat, p. 93. Quoted by Kolganov (1953), p. 368.
19. Rutkevich, M.N. (1985) 'O razvitii sovetskogo obshchestva k besklassovoi strukture', *Kommunist*, no. 18, pp. 37–8.
20. Emphasis in the original.
21. *Ekonomicheskaya Gazeta*, no. 9, 1986, p. 14.
22. Novye (1987) 'Novye yavleniya v lichnom podsobnom khozyaistve', *Voprosy Ekonomiki*, no. 7, p. 140.
23. Ibid., p. 141.
24. *Pravda*, 25 September 1987.
25. Wädekin (1973), p. 131.
26. *Trud*, 6 January 1979.
27. Dyachkov and Sorokin (1980), p. 66.
28. Grigorovskii, V.E. and Alekseev, M.A. (1968) *Lichnoe podsobnoe khozyaistvo kolkhoznikov, rabochikh sovkhozov i sluzhashchikh v SSSR*, Leningrad: LGU, p. 82.
29. *Trud*, 6 January 1979.
30. Voronin,V. (1980) 'Lichnye podsobnye khozyaistva i torgovlya', *Voprosy Ekonomiki*, no. 6, p. 119. The original source quoted here is Levin, A. and Nikitin, V. (1978) *Kolkhoznaya torgovlya v SSSR*. Moscow: Ekonomika, p. 61.
31. The seriousness of the matter is illustrated by a cartoon from *Krokodil*, which shows a row of stern-looking peasants selling vegetables at a market, and in front of them a man on his knees, hands clasped upwards to the sky. The latter is the brigade leader, trying desperately to get his *kolkhozniki* to come back in time to save the harvest (*Krokodil*, no. 30, 1977).
32. *Komsomolskaya Pravda*, 29 June 1983.
33. Kubyak, L.A. and Maksakova, G.P. (1981) 'Kollektivnoe sadovodstvo, ogorodnichestvo i zhivotnovodstvo rabochikh i sluzhashchikh', in Kalinkin, A.F. (ed.), *Lichnoe podsobnoe khozyaistvo, kollektivnoe sadovodstvo i ogorodnichestvo*, Moscow: Kolos, p. 139.
34. *Komsomolskaya Pravda*, 29 June 1983.
35. *Komsomolskaya Pravda*, 6 September 1983.

36. Ibid.
37. The wider aspects of corruption in the Soviet Union are dealt with in Simis, K. (1982) *USSR: The Corrupt Society; The Secret World of Soviet Capitalism*, New York: Simon and Schuster. See also Grossman, G. (1985) 'The second economy in the USSR and Eastern Europe: A bibliography', *Berkeley-Duke Occasional Papers on the Second Economy in the USSR*, no. 1, September.
38. Under Article 164 of the Penal Code.
39. *Trud*, 6 January 1979.
40. Konstitutsiya (1977) *Konstitutsiya (osnovnoi zakon) soyuza sotsialisticheskikh respublik*, Moscow: Politicheskaya Literatura.
41. 'O merakh po usileniyu borby s netrudovymi dokhodami' (*Pravda*, 28 May 1986).
42. 'Zakon ob individualnoi trudovoi deyatelnosti' (*Pravda*, 21 November 1986).
43. *Krokodil*, no. 23, 1987.
44. When speaking of activities that take place on or beyond the limits of the law, we must of course recognize that these are far from limited to agriculture. A very great deal of such activities also take place in the urban sector, in what is known as the 'second economy'. Problems in relation to the latter have been dealt with in a series of discussion papers edited by G. Grossman and V. Treml (*Berkeley-Duke Occasional Papers*).
45. *Sovetskaya Rossiya*, 12 May and 7 June 1985.
46. *Sovetskaya Rossiya*, 12 May 1985.
47. Ibid.
48. *Komsomolskaya Pravda*, 6 September 1983.
49. *Komsomolskaya Pravda*, 15 September 1983.
50. *Pravda*, 4 February 1980.
51. The agricultural tax on private plots in the RSFSR is only 85 kopeks to the *sotka* (a hundredth of a hectare), and even in places such as Uzbekistan where it may run up to 2.20 rubles, it is still considered low (Shmelev, G.I. (1985) 'Lichnoe podsobnoe khozyaistvo', *Novoe v zhizni nauke i tekhnike, Seriya Ekonomika*, no. 3, p. 17).

Chapter 5. The practice of harassment

1. *Literaturnaya Gazeta*, 3 June 1981, p. 10.
2. *Pravda*, 7 January 1982.
3. *Sotsialisticheskaya Industriya*, 1 October 1982.
4. FBIS, Daily Report: Soviet Union, 10 November 1986.
5. *Izvestia*, 6 December 1986.
6. Yanov, A. (1984) *The Drama of the Soviet 1960s. A Lost Reform*, Berkeley, CA: Institute of International Studies, p. 118.
7. The more general problems of relations between agriculture and its 'partners' are dealt with in greater detail in Hedlund, S. (1984) *Crisis in Soviet Agriculture*, London and Sydney: Croom Helm, New York: St Martin's, Chap. 6.

8. A real gem in this genre shows a large group of people camping in front of a building belonging to the (former) agricultural supply organization *Selkhoztekhnika*. On the door is a sign saying 'No Spare Parts', and it is obvious that people have waited for a long time, as some even have portable TV sets outside their tents. In the foreground is an official with a megaphone, saying that all those who have waited longer than three months must apply for a residence permit, a *propiska* (*Krokodil*, no. 8, 1983).
9. Shmelev, N. (1987) 'Avansy i dolgi', *Novyi Mir*, no. 6, p. 158.
10. According to one Soviet source, only in the years 1975–77 there occurred a doubling in the volume of machinery that was stripped in transit (*Pravda*, 26 October 1979).
11. Narkhoz (1985) *Narodnoe khozyaistvo SSSR v 1984 g*, Moscow: Finansy i statistika, pp. 216, 240; Narkhoz (1987) *Narodnoe khozyaistvo SSSR za 70 let: Yubileinyi statisticheskii ezhegodnik*, Moscow: Finansy i statistika, pp. 279, 281–2.
12. *Moscow News*, suppl., no. 4, 1988.
13. Palterovich, D. and Moskvin, S. (1982) 'Sredstva maloi mekhanisatsii dlya selskogo khozyaistva', *Voprosy Ekonomiki*, no. 8, p. 96.
14. Ibid., pp. 95–6.
15. Bush, K. (1967)'Mini-tractors for private plots advocated', *Radio Liberty Research*, CRD 115/67, 20 February.
16. *Izvestia*, 15 January 1977.
17. *Izvestia*, 15 August 1979.
18. *Pravda*, 8 December 1980.
19. *Pravda*, 29 December 1980.
20. TASS, 13 September 1981.
21. *Izvestia*, 23 January 1982.
22. Shmelev, G.I. (1984) *Podsobnye khozyaistva predpriyatii i naseleniyay*, Moscow: Znanie, p. 38.
23. See further *Radio Liberty Research*, RL 3/81, 31 December 1981, and RL 137/82, 24 March 1982.
24. Shmelev, G.I. (1985) 'Lichnoe podsobnoe khozyaistvo', *Novoe v zhizni, nauke i tekhnike*, Seriya Ekonomika, no. 3, p. 33.
25. *Izvestia*, 15 January 1977.
26. Western estimates of total labour use in the private sector show that since the 1950s this has been rather stable at around 10–12 million man-years (Goodman, A., Hughes, M. and Schroeder, G. (1987) 'Raising the efficiency of Soviet farm labor: problems and prospects', in US Congress (1987) Joint Economic Committee, *Gorbachev's Economic Plans*, vol. II, Washington, DC: Government Printing Office, p. 106). This comes to about 3.6–4.4 billion man-days, which corresponds fairly well with the Soviet figures given here.
27. Shmelev, G.I. (1983) *Lichnoe podsobnoe khozyaistvo: vozmozhnosti i perspektivy*, Moscow: Politicheskaya Literatura, p. 45.
28. Ibid., p. 43.
29. *Sovetskaya Torgovlya*, 14 October 1982.

30. *Sovetskaya Torgovlya*, 28 October 1982.
31. *Radio Liberty Research*, RL 480/82, 30 November 1982.
32. Shmelev (1983), p. 44.
33. *Radio Liberty Research*, RL 480/82, 30 November 1982.
34. *Selskaya Zhizn*, 4 February 1981.
35. *Sovetskaya Torgovlya*, 14 October 1982.
36. See further a collection of such letters in *Radio Liberty Research*, 480/82, 30 November 1982.
37. Shmelev (1983), p. 44.
38. *Komsomolskaya Pravda*, 15 May 1981.
39. *Komsomolskaya Pravda*, 9 August 1981.
40. Shmelev (1983), p. 46.
41. The total number of horses in the Soviet Union has been reduced from 21.1 million in 1941 to 9.9 million in 1961 and 5.8 million in 1985 (Narkhoz (1986), p. 236). In addition, one might perhaps reasonably assume that many of those remaining will be intended for the highly popular sport of horse racing, rather than for hard work on the farms. For some odd reason the 1987 edition of *Narodnoe khozyaistvo* includes figures on horses for 1917–1940, but not for the subsequent period (Narkhoz (1987), p. 253).
42. Kuznetsova, T. (1984) 'Resursnoe obespechenie lichnogo podsobnogo khozyaistva', *Voprosy Ekonomiki*, no. 11, p. 100.
43. Ibid., p. 99.
44. In round figures, the private sector's share of major livestock holdings was reduced by about a third, whereas the reduction in absolute terms came to about a fifth for cows, and a fourth for hogs (Narkhoz (1976), pp. 391–2, Narkhoz (1981), p. 245, Narkhoz (1986), p. 236).
45. Shmelev (1985), p. 15.
46. Sidorenko, V.I. (1985) 'Vazhnyi istochnik popolneniya prodovolstvennogo fonda', *Planirovanie i uchet v selskokhozyaistvennykh predpriyatiyakh*, no. 3, p. 4.
47. Shmelev (1983), pp. 53–4.
48. Prodovolstvennaya (1982) *Prodovolstvennaya programma SSSR na period do 1990 goda i mery po ee realizatsii. Materialy maiskogo plenuma TsK KPSS*, Moscow: Politizdat, pp. 98–9.
49. *Izvestia*, 19 April 1982.
50. The *magnum opus* on the *kolkhoz* markets is Kerblay, B. (1968) *Les marchés paysans en URSS*, Paris: Mouton. See also Wädekin, K.-E. (1973) *The Private Sector in Soviet Agriculture*, Berkeley, CA: University of California Press, Chap. 6.
51. The right to trade there at freely formed prices is granted in Article 40 of the Soviet Civil Code (Wädekin (1973), p. 362).
52. *Trud*, 6 January 1979.
53. Narkhoz (1987), p. 484.
54. Narkhoz (1976), p. 617, and Narkhoz (1981), p. 425.
55. Narkhoz (1987), p. 455.
56. Shmelev (1983), pp. 67–8. See also Wädekin, K.-E. (1985) 'The role

of the kolkhoz market: A quantitative assessment', *Radio Liberty Research*, RL 155/85, 13 May.
57. Ofer, G. and Vinokur, A. (1980) 'Private sources of income of the Soviet urban household', *RAND Report*, R-2359-NA, August.
58. Based on a survey of 1,000 recent emigrés, Treml, V.D. (1985) ('Purchases of food from private sources in Soviet urban areas', *Berkeley-Duke Occasional Papers on the Second Economy in the USSR*, no. 3, p. 8) claims that actual sales on urban *kolkhoz* markets in 1977 were 'about 5.6 times higher than reported in official Soviet statistical sources.' This places him at the top of the range of such estimates.
59. Shenfield, S. (1984) 'How reliable are published Soviet data on the *kolkhoz* markets?', *CREES Discussion Papers, General Series*, Gl, November, p. 35.
60. Narkhoz (1987), p. 455.
61. This view is derived by Shenfield (1984), p. 46, from personal conversation with Professor V. Shvyrkov, formerly of the Gosplan and now at San Francisco University.
62. Narkhoz (1987), p. 455.
63. Ibid., p. 485.
64. Treml (1985), p. 17. On methodology, see also Shenfield (1984) *passim*, and Severin, B.S. (1979) 'USSR: The all-union and Moscow collective farm market price indexes', *ACES Bulletin*, 21, no. 1.
65. Shmelev (1983), p. 74.
66. Ibid., pp. 70–1.
67. *Trud*, 6 January 1979.
68. Ibid.
69. *Literaturnaya Gazeta*, 6 May 1981, p. 12.
70. Narkhoz (1987), p. 484.
71. Kaplan, C.S. (1987) *The Party and Agricultural Crisis Management in the USSR*, Ithaca and London: Cornell University Press.
72. Shmelev (1983), p. 58.
73. Wädekin, K.-E. (1986) 'Private Leistungen für den Lebensmittelsmarkt der UdSSR', *Osteuropa* vol. 36, no. 1, p. 55.
74. See further ibid., pp. 57–62.
75. Shmelev (1983), p. 60. Over time, the share of potatoes and vegetables has been increasing (up from 30 to 34 per cent respectively, in 1975, to 45 per cent in 1980), while meat and milk have been decreasing (down from 67 and 97 per cent in 1975, to 35 and 94 per cent in 1980) (Shmelev (1985), pp. 21–2).
76. Voronin, V. (1980) 'Lichnye podsobnye khozyaistva i torgovlya', *Voprosy Ekonomiki*, no. 6, pp. 120–1.
77. Peredovaya (1982) 'Peredovaya', *Kommunist*, no. 11, p. 7. The nature and causes of such losses are discussed in Flynn, J. and Severin, B. S. (1987) 'Soviet agricultural transport: Bottlenecks to continue', appendix B, in US Congress (1982) Joint Economic Committee, *Soviet Economy in the 1980s: Problems and Prospects*, vol. II, Washington, DC: Government Printing Office.

78. *Pravda*, 26 February 1986.
79. Novye (1987) 'Novye yavleniya v lichnom podsobnom khozyaistve', *Voprosy Ekonomiki*, no. 7, p. 142.
80. Ibid.
81. In 1982, a model charter, *Tipovoe polozhenie buro torgovykh uslug*, was presented to regulate the activities of these bureaus, establishing *inter alia* that the vice-director of the market should normally be in charge of the trade bureau (Shmelev (1985), p. 24).
82. *Trud*, 6 January 1979.
83. Voronin (1980), p. 123.
84. Shmelev (1985), pp. 25–6.
85. *Pravda*, 5 March 1986.
86. These problems are discussed in great detail in Flynn and Severin (1987).
87. *Pravda*, 14 July 1986.
88. *Izvestia*, 24 September 1981.
89. *Izvestia*, 27 September 1981.
90. *Izvestia*, 28 June 1982.
91. *Izvestia* 27 September 1981.
92. Shmelev (1983), p. 82.
93. Ibid., p. 79.
94. *Pravda*, 10 September 1983.
95. *Pravda*, 14 July 1986.
96. Novye (1987), p. 141.
97. Ibid., p. 142.
98. Ibid., p. 144.
99. Ibid., p. 146.
100. Ibid., p. 147.
101. Shmelev (1983), p. 16.
102. Ibid., p. 28.
103. *Izvestia*, 27 September 1981.

Chapter 6. Response to decline

1. Hirschman, A. O. (1970) *Exit, Voice and Loyalty. Responses to Decline in Firms, Organizations and States*, Cambridge, MA and London: Harvard University Press.
2. Ibid., p. 1.
3. See further Williamson, O. (1976) 'Some uses of the Exit-Voice Approach – Discussion', *American Economic Review* 66, no. 2.
4. Hirschman (1970), p. 26.
5. See further ibid., Chap. 7.
6. Other examples of such use of economic models in political science are Downs, A. (1957) *An Economic Theory of Democracy*, New York: Harper & Row, and Olson, M. (1965) *The Logic of Collective Action*, Cambridge, MA and London: Harvard University Press.
7. Hirschman (1970), p. 19.
8. Ibid., pp. 44–5.

9. See Hirschman, A.O. (1986) 'Exit and Voice: An expanding sphere of influence', in: *Rival Views of Market Society and Other Recent Essays*, New York: Viking, pp. 85–99, for further reference.
10. Hirschman (1970), p. 44.
11. Birch, A.H. (1975) 'Economic models in political science: the case of "Exit, Voice and Loyalty"', *British Journal of Political Science* 5, no. 2, pp. 74–5; Laver, M. (1976) '"Exit, Voice and Loyalty" revisited: the strategic production and consumption of public and private goods', *British Journal of Political Science* 6, no. 4, p. 477.
12. Barry, B. (1974) 'Review article: "Exit, Voice and Loyalty"', *British Journal of Political Science* 4, no. 1, p. 95.
13. See further Elster, J. (1983) *Sour Grapes: Studies in the Subversion of Rationality*, Cambridge: Cambridge University Press, Chap. II.
14. White S. (1986) 'Economic performance and communist legitimacy', *World Politics* 38, no. 3, p. 470.
15. Ibid., pp. 471–81.
16. This concept was originally introduced in Hedlund, S. (1987) 'Soft options in central control', in: Hedlund, S. (ed.), *Incentives and Economic Systems: Proceedings of the eighth Arne Ryde Symposium, Frostavallen, 26–7 Aug. 1987*, London and Sydney: Croom Helm.
17. I owe this distinction to Alec Nove.
18. The expression has been borrowed from Yanov, who argues that the imposition of the *kolkhoz* system has led to a '"de-peasantization" of the peasantry by destroying both their incentive and their time-honoured habits of work on the land' (Yanov, A. (1984) *The Drama of the Soviet 1960s. A Lost Reform*, Berkeley, CA: Institute of International Studies, p. 22).
19. *Komsomolskaya Pravda*, 21 August 1982.
20. Vasilev I. (1982) 'Vozvrashchenie k zemle', *Nash sovremennik*, no. 6, p. 7.
21. Lewin, M. (1985) *The Making of the Soviet System. Essays in the Social History of Interwar Russia*, New York: Pantheon Books, p. 268.
22. Conquest, R. (1986) *The Harvest of Sorrow: Soviet Collectivization and the Terror Famine*, New York and Oxford: Oxford University Press, p. 143.
23. Yanov (1984), p. 5. As so often is the case in matters regarding the Soviet Union, we must distinguish here between the formal and the actual. In actual practice, Soviet peasants have of course found a number of different ways to leave for the cities, but it remains a fact that until recently they were legally treated as second-rate citizens, with all the social and psychological consequences that this must have had.
24. Section II, Article 7, of the model charter.
25. Yanov (1984), p. 40. The following condemnation is offered: 'The *kolkhoz* countryside, an invention of Stalin, the artificial fruit of dictatorship and the result of the destruction of the peasant elite, really cannot be left to its own devices. For it will cease to work (even as badly as it works now) as soon as the club is laid down' (ibid., p. 56).

26. Shmelev, G.I. (1983) *Lichnoe podsobnoe khozyaistvo: vozmozhnosti i perspektivy*. Moscow: Politicheskaya Literatura, p. 24.
27. Vasilev (1982) p. 7.
28. See further Hedlund, S. (1984) *Crisis in Soviet Agriculture*, London and Sydney: Croom Helm, New York: St. Martin's, p. 81.
29. Vasilev (1982), p. 8.
30. Ibid., p. 12.
31. *Literaturnaya Gazeta*, 10 February 1982, p. 13.
32. Vasilev (1982), p. 10.
33. Shmelev (1983), p. 74.
34. Dyachkov and Sorokin, A. (1980) 'Rol lichnogo podsobnogo khozyaistva', *Ekonomika selskogo khozyaistva*, no. 1, p. 65. The cited source is Mashenkov, V.F. (1965) *Ispolzovanie trudovykh resursov selskoi mestnosti*, Moscow: Ekonomika, p. 55.
35. Lewin (1985), p. 186.
36. Shmelev (1983), p. 26.
37. Ibid., p. 37.
38. Chernichenko, Yu. (1978) 'Pro kartoshku', *Nash sovremennik*, no. 6, p. 138.
39. See e.g. 'O dopolnitelnykh merakh po obespecheniyu uborki urozhaya, zagotovok selskokhozyaistvennykh produktov i kormov v 1983 godu i uspeshnogo provedeniya zimovki skota v period 1983/84 goda' (*Izvestia*, 21 May 1983). Although the labour statistics quoted above certainly present a rather gloomy picture of the costs that are involved here, we may add that the burden is not limited to labour alone. On a permanent basis, agriculture accounts for some 20 per cent of the total number of trucks that can be found in the Soviet economy as a whole. In absolute numbers, this comes to about 1.8 million, which no doubt is a rather impressive figure. Every year during harvest time, however, an additional 700,000–800,000 trucks are requisitioned, trucks which obviously must be taken from other productive uses (Flynn, J. and Severin, B.S. (1987) 'Soviet agriculture transport bottlenecks to continue', in: US Congress (1987) Joint Economic Committee, *Gorbachev's Economic Plans*, vol. II, Washington, DC: Government Printing Office, pp. 65–6).
40. A collection of such evidence is published in *Radio Liberty Research*, RL 269/83, 18 July 1983.
41. In a series of interesting articles, Kapitolina Kozhevnikova, correspondent at *Literaturnaya Gazeta*, has presented and discussed a range of Soviet views on the desirability of such activities. See *Literaturnaya Gazeta*, 11 January, 8 February and 26 July 1978.
42. *Pravda*, 24 May 1980.
43. Ibid.
44. See further Wädekin, K.-E. (1985) ('The private agricultural sector in the 1980s', *Radio Liberty Research*, RL 251/85, 2 August) who gives the effect as 'minimal'.
45. Wädekin, K.-E. (1983) 'The impact of official policy on the number

of livestock in the Soviet private farming sector', *Radio Liberty Research*, RL 224/83, 9 June.

46. *Pravda*, 25 September 1987.
47. Hedlund, S. (1989) 'Exit, Voice and Loyalty – Soviet style', *Coexistence* 26, no. 2.
48. Hirschman, A.O. (1981) *Essays in Trespassing: Economics to Politics and Beyond*, Cambridge: Cambridge University Press, p. 227.
49. Timofeev, L. (1985) *Soviet Peasants (or: The Peasants Art of Starving)*, New York: Telos; Humphrey, C. (1983) *Karl Marx Collective: Economy, Society and Religion in a Siberian Collective Farm*, Cambridge: Cambridge University Press.
50. Wells, H.G. (1920) *Russia in the Shadows*, London: Stodder & Houghton, pp. 136–7.
51. Churchill, W.S. (1950) *World War II*, vol. 4. Boston, p. 498.
52. *Pravda*, 7 August 1987.
53. Shmelev, G.I. (1987) *Semya beret podryad*, Moscow: Agropromizdat, pp. 15–16.

Chapter 7. A Pyrrhic victory and its consequences

1. *Pravda*, 26 February 1986.
2. One might of course even argue that productive Exit in agriculture represents a net gain, and thus is not costly at all. This argument, however, is in some sense a fallacious one, since it rests on accepting that the socialized sector *must* function so poorly that the opportunity cost of resources withdrawn from it is necessarily very low. It does not recognize that the poor functioning of that sector might actually to some extent be a consequence of the Exit.
3. Sobranie (1987) *Sobranie postanovlenii pravitelstva SSSR*, no. 10, pp. 195–222; *Izvestia*, 6 March 1988. For comments, see Hanson, P. (1988) 'The draft law on cooperatives: an assessment', *Radio Liberty Research*, RL 111/88, 15 March.
4. Hanson (1988), p. 2. Reference is made to a survey of letters in *Izvestia*, 27 February 1988.
5. See further Gerner, K. and Hedlund, S. (1988) 'Die polnische Dauerkrise', *Osteuropa* 38, no. 5.
6. Laba, R. (1986) 'Worker roots of solidarity', *Problems of Communism*, vol. 35 no. 7, p. 65.
7. Motyl, A. (1987) *Will the Non-Russians Rebel? State, Ethnicity and Stability in the USSR*, Ithaca and London: Cornell University Press, pp. 107–10.
8. Ibid., pp. 117–18.
9. This concept was first introduced and discussed in Hedlund, S. (1987) 'Soft options in central control', in Hedlund, S. (ed.), *Incentives and Economic Systems: Proceedings of the Eighth Arne Ryde Symposium, Frostavallen, 26–7 Aug. 1987*, London and Sydney: Croom Helm.
10. See further Yanov, A. (1984) *The Drama of the Soviet 1960s. A*

Lost Reform, Berkeley, CA: Institute of International Studies.

11. The story was originally published in *Literaturnaya Gazeta*, 7 August 1968. It also appears in Yanov (1984), pp. 19–20.
12. On the problems of individual and collective rationality, see further Elster, J. (1983) *Sour Grapes: Studies in the Subversion of Rationality*, Cambridge: Cambridge University Press, Chap. 1. The Prisoner's Dilemma is discussed *inter alia* in Rapoport, A. (1982) 'Prisoner's Dilemma – recollections and observations', in Barry, B. and Hardin, R. (eds), *Rational Man and Irrational Society*, Beverly Hills, London and New Delhi: Sage.
13. Motyl (1987), p. 71.
14. North, D. (1981) *Structure and Change in Economic History*, New York and London: Norton, p. 45.
15. Ibid., p. 47.
16. Ibid. Elster has noted the same, albeit from a different angle, in pointing out that it may sometimes be necessary to delude individuals out of making their own narrowly rational decisions: 'Irrationality rather than duty may be the cement of society, a socially beneficial illusion, like Voltaire's God' (Elster, J. (1985) 'Rationality, morality and collective action', *Ethics* 96, no. 1, p. 146).
17. North (1981), p. 53.
18. Ibid., p. 49.
19. *Selskaya Zhizn*, 4 January 1985.
20. Lewin, M. (1985) *The Making of the Soviet System. Essays in the Social History of Interwar Russia*, New York: Pantheon Books, pp. 186–7.
21. Ibid., p. 187.
22. We might mention in this context a highly interesting correspondence between Karl Marx and the Russian *narodnik* Vera Zasulich, regarding whether the *mir* might be used as a basis for building a socialist agriculture without going through the phase of capitalist transformation. The essence of Marx's reply was that his theory was conceived for a Western European context, and could thus not be directly applied to Russia. The relevance of this example to our presentation lies in the fact that although their strength may be subject to controversy, the Russian *mir* did feature elements of a collective nature that are largely absent from the Soviet *kolkhozy* and *sovkhozy* of today. On the famous correspondence, see further Mitrany, D. (1951) *Marx Against the Peasant: A Study in Social Dogmatism*, Chapel Hill, NC: University of North Carolina Press, pp. 31–3.
23. Quoted by Robinson, G.T. (1961) *Rural Russia Under the Old Regime*. London and New York: Longman, p. 194.
24. E.g. Volin, L. (1970) *A Century of Russian Agriculture*, Cambridge, MA: Harvard University Press, p. 105.
25. Lewin (1985), p. 187.
26. Vasilev, I. (1982) 'Vozvrashchenie k zemle', *Nash sovremennik*, no. 6, pp. 13–14.

27. Hirschman, A.O. (1970) *Exit, Voice and Loyalty. Responses to Decline in Firms, Organizations and States.* Cambridge, MA and London: Harvard University Press, p. 55.
28. See further Kornai, J. (1979) 'Resource-constrained versus demand-constrained systems', *Econometrica* 47, no. 4.
29. Vasilev (1982), p. 10.
30. *Selskaya Zhizn,* 17 May 1984.
31. Yanov (1984), p. 87 and Chap. 3
32. Ibid., p. 19.
33. *Pravda,* 3 October 1987.
34. *Pravda,* 7 August 1987.
35. According to Narkhoz (1987), p. 300, the 'apparatus of controlling organs' in agriculture employed in 1980 a total of 0.44 million people. By 1985, that figure had risen to 0.51 million. Under the impact of *perestroika,* however, this trend of growth has been broken. For 1986, no more than 0.47 million was recorded.
36. *Literaturnaya Gazeta,* 11 May 1988, p. 2.
37. Shmelev, G.I. (1985) 'Semya beret podryad', *Selskaya Nov,* no. 10, p. 9.
38. See Wädekin, K.-E. (1984) '"Contract" and "Normless" labour on Soviet farms', *Radio Liberty Research,* RL 49/84, 8 February.
39. See further Pospielovsky, D. (1970) 'The "link system" in Soviet agriculture', *Soviet Studies* 21, no. 4.
40. Shmelev, G.I. (1985)'Lichnoe podsobnoe khozyaistvo', *Novoe v zhizni, nauke i tekhnike, Seriya Ekonomika,* no. 3, pp. 5–6.
41. *Pravda,* 25 May 1982.
42. *Pravda,* 19 March 1983.
43. *Ekonomicheskaya Gazeta,* no. 41, 1985, pp. 11–13.
44. *Selskaya Zhizn,* 25 January 1987.
45. *Pravda,* 25 September 1987.
46. Shmelev, G.I. (1987) 'Semya beret podryad', Moscow: Agroporomizdat, p. 7.
47. Laird, R.D. and Laird, B.A. (1987) 'Perestroika in agriculture: Gorbachev's rural "revolution"?', paper presented to the Eighth International Conference on Soviet and East European Agriculture, Berkeley, August 1987.
48. Quoted by Goodman, A., Hughes, M. and Schroeder, G. (1987) 'Raising the efficiency of Soviet farm labor: problems and prospects', in US Congress (1987) Joint Economic Committee, *Gorbachev's Economic Plans,* vol. II, Washington, DC : Government Printing Office, p. 120.
49. Gagnon, V.P. (1987) 'Gorbachev and the collective contract brigade', *Soviet Studies* 39, no. 1, p. 12.
50. Wädekin (1984), p. 4.
51. Ibid., p. 5.
52. Ironically, it was precisely the former kind of cooperatives that Lenin had in mind in writing one of his very last pamphlets, *On Cooperation,* published in *Pravda* on 26 and 27 May 1923.

53. Quoted by Nahaylo, B. (1987)'Interview with Tatyana Zaslavskaya', *Radio Liberty Research*, RL 365/87, 15 September.
54. On the *kibbutz*, see Barkai, H. (1987) 'Kibbutz efficiency and the incentive conundrum', in Hedlund, S.(ed.), *Incentives and Economic Systems: Proceedings of the Eighth Arne Ryde Symposium, Frostvallen, 26–27 August 1987*, London and Sydney: Croom Helm; and on the broader aspects of socialized agricultural labour, Hedlund, S. (1985) 'On the socialization of labour in rural cooperation', in Lundahl, M. (ed.), *The Primary Sector in Economic Development: Proceedings of the Seventh Arne Ryde Symposium, Frostavallen. 29–30 Aug. 1985*, London and Sydney: Croom Helm.
55. *Moscow News*, suppl., no. 4, 1988.
56. *Literaturnaya Gazeta*, no. 32, 1987, p. 10. No less than 15 agricultural economists, charged and convicted in the early 1930s, were rehabilitated at this time, including Aleksandr Chayanov, Nikolai Kondratiev and Aleksandr Chelintsev.
57. See Thorner, D., Kerblay, B. and Smith, R.E.F. (eds.) (1966) *A.V.Chayanov on the Theory of Peasant Economy*, Homewood, IL: Irwin, pp. xliv–xlvi.
58. *Literaturnaya Gazeta*, 3 June 1981, p. 10.

Bibliography

Barkai, Haim (1987) 'Kibbutz efficiency and the incentive conundrum', in S. Hedlund (ed.) *Incentives and Economic Systems: Proceedings of the Eighth Arne Ryde Symposium, Frostavallen, 26-77 Aug.*, London and Sydney: Croom Helm.
Barry, Brian (1974) 'Review article: "Exit, Voice and Loyalty"', *British Journal of Political Science* 4, no. 1.
Belyanov, Vladislav A. (1970) *Lichnoe podsobnoe khozyaistvo pri sotsializme*, Moscow: Nauka.
Birch, A. H. (1975) 'Economic models in political science: the case of "Exit, Voice and Loyalty"', *British Journal of Political Science* 5, no. 2.
Blyakhman, Leonid S. and Shkaratan, Ovsei I. (1973) *NTR, rabochii klass, intelligentsiya*, Moscow: Politicheskaya Literatura.
Bradley, Michael E. (1971) 'Incentives and labor supply on Soviet collective farms', *Canadian Journal of Economics* 4, no. 3.
Bush, Keith (1967) 'Mini-tractors for private plots advocated', *Radio Liberty Research*, CRD 115/67, 20 February.
Chernichenko, Yu. (1978) 'Pro kartoshku', *Nash sovremennik*, no. 6
Churchill, Winston (1950) *The Second World War*, Vol. 4, Boston: Houghton Mifflin.
Conquest, Robert (1986) *The Harvest of Sorrow: Soviet Collectivization and the Terror Famine*, New York and Oxford: Oxford University Press.
Doolittle, Penelope and Hughes, Margaret (1987) 'Gorbachev's agricultural policy: building on the Brezhnev Food Program', in US Congress, Joint Economic Committee, *Gorbachev's Economic Plans*, vol. II, Washington, DC: Government Printing Office.
Downs, Anthony (1957) *An Economic Theory of Democracy*, New York: Harper and Row.
Dyachkov, G. and Sorokin, A. (1980) 'Rol lichnogo podsobnogo khozyaistva', *Ekonomika selskogo khozyaistva*, no. 1.
Elster, Jon (1983) *Sour Grapes: Studies in the Subversion of Rationality*, Cambridge: Cambridge University Press.
Elster, Jon (1985) 'Rationality, morality and collective action', *Ethics* 96, no. 1
Flynn, Judith and Severin, Barbara S. (1987) 'Soviet agricultural transport:

bottlenecks to continue', in US Congress, Joint Economic Committee, *Gorbachev's Economic Plans*, vol. II, Washington DC: Government Printing Office.
Gagnon, V. P. (1987) 'Gorbachev and the collective contract brigade', *Soviet Studies* 39, no. 1
Gerner, Kristian and Hedlund, Stefan (1988) 'Die polnische Dauerkrise', *Osteuropa* 38, no. 5.
Goodman, Ann, Hughes, Margaret and Schroeder, Gertrude (1987) 'Raising the efficiency of Soviet farm labor: problems and prospects', in US Congress, Joint Economic Committee, *Gorbachev's Economic Plans*, vol. II, Washington, DC: Government Printing Office.
Gray, Kenneth R. (1987) 'Reform and resource allocation in Soviet agriculture', in US Congress, Joint Economic Committee, *Gorbachev's Economic Plans*, vol. II, Washington, DC: Government Printing Office.
Grigorovskii, Vasilii E. and Alekseev, Mikhail A. (1968) *Lichnoe podsobnoe khozyaistvo kolkhoznikov, rabochikh sovkhozov i sluzhashchikh v SSSR*, Leningrad: LGU.
Grossman, Gregory (1985) 'The second economy in the USSR and Eastern Europe: a bibliography', *Berkeley-Duke Occasional Papers on the Second Economy in the USSR*, no. 1, September.
Hanson, Philip (1988a) 'Gorbachev addresses kolkhozniks' congress', *Radio Liberty Research*, 131/88, 24 March.
Hanson, Philip (1988b), 'The draft law on cooperatives: an assessment', *Radio Liberty Research*, RL 111/88, 15 March.
Hedlund, Stefan (1984) *Crisis in Soviet Agriculture*, London and Sydney: Croom Helm, New York: St. Martin's.
Hedlund, Stefan (1985) 'On the socialization of labor in rural cooperation', in Mats Lundahl, (ed.), *The Primary Sector in Economic Development: Proceedings of the Seventh Arne Ryde Symposium, Frostavallen, 29–30 Aug. 1985*, London and Sydney: Croom Helm.
Hedlund, Stefan (1987) 'Soft options in central control', in S. Hedlund (ed.) *Incentives and Economic Systems: Proceedings of the Eighth Arne Ryde Symposium, Frostavallen, August 26-27, 1987*, London and Sydney: Croom Helm.
Hedlund, Stefan (1989) 'Exit, Voice and Loyalty – Soviet style', *Coexistence* 26, no. 2.
Hirschman, Albert O. (1970) *Exit, Voice and Loyalty. Responses to Decline in Firms, Organizations and States*, Cambridge, MA and London: Harvard University Press
Hirschman, Albert O. (1981) *Essays in Trespassing: Economics to Politics and Beyond*, Cambridge: Cambridge University Press.
Hirschman, Albert O. (1986) 'Exit and Voice: an expanding sphere of influence', in *Rival Views of Market Society, and Other Recent Essays*, New York: Viking.
Humphrey, Caroline (1983) *Karl Marx Collective: Economy, Society and Religion in a Siberian Collective Farm*, Cambridge: Cambridge University Press.
Ioffe, M. (1982) 'Podsobnye khozyaistva promyshlennykh predpriyatii i

organisatsii', *Planovoe khozyaistvo*, no. 11
Jasny, Naum (1951) '*Kolkhozy*, the Achilles' heel of the Soviet regime', *Soviet Studies* 3, no. 2.
Johnson, D. Gale and McConnell Brooks, Karen (1983) *Prospects for Soviet Agriculture in the 1980s*, Bloomington, IN: Indiana University Press.
Kalinkin, Aksenii F. (ed.)(1981) *Lichnoe podsobnoe khozyaistvo, kollektivnoe sadovodstvo i ogorodnichestvo*, Moscow: Kolos.
Kalinkin, Aksenii F. (1982) 'Razvitie lichnogo podsobnogo khozyaistva', *Ekonomika selskogo khozyaistva*, no. 4.
Kaplan, Cynthia S. (1987) *The Party and Agricultural Crisis Management in the USSR*, Ithaca and London: Cornell University Press.
Karcz, Jerzy (1966) 'Seven years on the farm: retrospect and prospects', in US Congress, Subcommittee on Foreign Economic Policy, *New Directions in the Soviet Economy*, Washington, DC: Government Printing Office.
Karakhanova, T. M. and Patrushev, V. D. (1983) 'Kollektivnoe ogorodnichestvo i sadovodstvo – reserv prodovolstvennogo obespecheniya', *Sotsiologicheskie Issledovaniya*, no. 2.
Kerblay, Basile (1968) *Les marchés paysans en URSS*, Paris: Mouton.
Khronika (1982), 'Khronika nauchnoi zhizni: Problemy lichnogo podsobnogo khozyaistva', *Izvestiya Akademii Nauk SSSR. Seriya ekonomicheskaya*, no. 3.
Kolganov, M. V. (1953) *Sobstvennost v sotsialisticheskom obshchestve*, Moscow: AN SSSR.
Konstitutsiya (1961) *Konstitutsiya (osnovnoi zakon) soyuza sotsialisticheskikh respublik*, Moscow: Politicheskaya Literatura.
Konstitutsiya (1977) *Konstitutsiya (osnovnoi zakon) soyuza sotsialisticheskikh respublik*, Moscow: Politicheskaya Literatura.
Kornai, János (1979) 'Resource-constrained versus demand-constrained systems', *Econometrica* 47, no. 4
Kozyr, M. I. (1981) 'Pravovye osnovy vedeniya lichnogo podsobnogo khozyaistva', in A.F. Kalinkin (ed.) *Lichnoe podsobnoe khozyaistvo, kollektivnoe sadovodstvo i ogorodnichestvo*, Moscow: Kolos.
Kubyak, L. A. and Maksakova, G. P. (1981) 'Kollektivnoe sadovodstvo, ogorodnichestvo i zhivotnovodstvo rabochikh i sluzhashchikh', in A.F. Kalinkin (ed.) *Lichnoe podsobnoe khozyaistvo, kollektivnoe sadovodstvo i ogorodnichestvo*, Moscow: Kolos.
Kutin, Evegenii M. (1974) *Kollektivnoe sadovodstvo i ogorodnichestvo*, Moscow: Yuridicheskaya Literatura.
Kuznetsova, T. (1984) 'Resursnoe obespechenie lichnogo podsobnogo khozyaistva', *Voprosy Ekonomiki*, no. 11.
Laba, Roman (1986) 'Worker roots of Solidarity', *Problems of Communism* 35, no. 7.
Laird, Roy D. and Crowley, Edward (eds) (1965) *Soviet Agriculture: The Permanent Crisis*, New York: Praeger.
Laird, Roy D. and Laird, Betty A. (1987) 'Perestroika in agriculture: Gorbachev's rural "revolution"?', in paper presented to the Eighth

International Conference on Soviet and East European Agriculture, Berkeley, August 1987.
Lane, Ann (1982) 'USSR: private agriculture on center stage', in US Congress, Joint Economic Committee, *Soviet Economy in the 1980s: Problems and Prospects*, vol. II, Washington, DC: Government Printing Office.
Larionova, K. (ed.) (1963) *Politicheskaya ekonomiya. Uchebnoe posobie*, Moscow.
Laver, Michael (1976) '"Exit, Voice and Loyalty" revisited: the strategic production and consumption of public and private goods', *British Journal of Political Science* 6, no. 4.
Leninskaya (1978) *Leninskaya agrarnaya politika KPSS: sbornik vazhneishikh dokumentov (mart 1965 g. – ijul 1978 g.)*, Moscow: Politizdat.
Leninskaya (1983) *Leninskaya agrarnaya politika KPSS: sbornik vazhneishikh dokumentov (avgust 1978 g. - avgust 1982 g.)*, Moscow: Politizdat.
Levin, A. and Nikitin, V. (1978) *Kolkhoznaya torgovlya v SSSR*, Moscow: Ekonomika.
Lewin, Moshe (1965), 'The immediate background to Soviet collectivization', *Soviet Studies* 17, no. 2.
Lewin, Moshe (1966) 'Who was the Soviet kulak?', *Soviet Studies* 18, no. 2.
Lewin, Moshe (1968) *Russian Peasants and Soviet Power: A Study of Collectivization*, London: Allen and Unwin.
Lewin, Moshe (1974) *Political Undercurrents in Soviet Economic Debates: From Bukharin to the Modern Reformers*, Princeton: Princeton University Press.
Lewin, Moshe (1985) *The Making of the Soviet System. Essays in the Social History of Interwar Russia*, New York: Pantheon Books.
Malish, Anton (1982) 'The Food Program: a new policy or more rhetoric?', in US Congress, Joint Economic Committee, *Soviet Economy in the 1980s: Problems and Prospects*, vol. II, Washington, DC: Government Printing Office.
Manevich, E. (1981) 'Ratsionalnoe ispolzovanie rabochei sily', *Voprosy Ekonomiki*, no. 9.
Marx, Karl and Engels, Friedrich (1977) *Manifesto of the Communist Party*, Moscow: Progress.
Mashenkov, Vladimir F. (1965) *Ispolzovanie trudovykh resursov selskoi mestnosti*, Moscow: Ekonomika.
McCauley, Martin (1976) *Khrushchev and the Development of Soviet Agriculture: The Virgin Lands Programme, 1953–64*, London: Macmillan.
Mitrany, David (1951) *Marx Against the Peasant: A Study in Social Dogmatism*, Chapel Hill, NC: University of North Carolina Press.
Motyl, Alexander (1987) *Will the Non-Russians Rebel? State, Ethnicity, and Stability in the USSR*, Ithaca and London: Cornell University Press.

Nahaylo, Bohdan (1987) 'Interview with Tatyana Zaslavskaya', *Radio Liberty Research*, RL 365/87, 15 September.
Narkhoz (1963) *Narodnoe khozyaistvo SSSR v 1962 g.*, Moscow: Finansy i statistika.
Narkhoz (1966) *Narodnoe khozyaistvo SSSR v 1965 g.*, Moscow: Finansy i statistika.
Narkhoz (1968) *Narodnoe khozyaistvo SSSR v 1967 g.*, Moscow: Finansy i statistika.
Narkhoz (1976) *Narodnoe khozyaistvo SSSR v 1975 g.*, Moscow: Finansy i statistika.
Narkhoz (1981) *Narodnoe khozyaistvo SSSR v 1980 g.*, Moscow: Finansy i statistika.
Narkhoz (1985) *Narodnoe khozyaistvo SSSR v 1984 g.*, Moscow: Finansy i statistika.
Narkhoz (1986) *Narodnoe khozyaistvo SSSR v 1985 g.*, Moscow: Finansy i statistika.
Narkhoz (1987) *Narodnoe khozyaistvo SSSR za 70 let: Yubileinyi statisticheskii ezhegodnik*, Moscow: Finansy i statistika.
North, Douglass (1981) *Structure and Change in Economic History*, New York and London: Norton.
Nove, Alec (1977) *The Soviet Economic System*, London: Allen and Unwin.
Novoe (1969) *Novoe v zemelnom zakonodatelstve*, Moscow: Yuridicheskaya Literatura.
Novye (1987) 'Novye yavleniya v lichnom podsobnom khozyaistve', *Voprosy Ekonomiki*, no. 7.
Ofer, Gur and Vinokur, Aron (1980) 'Private sources of income of the Soviet urban household', *RAND Report*, R-2359-NA, August.
Olson, Mancur (1965) *The Logic of Collective Action*, Cambridge, MA. and London: Harvard University Press.
Ostrovskii, V. B. (1967) *Kolkhoznoe krestyanstvo SSSR. Politika partii v derevne i ee sotsialno-ekonomicheskie resultaty*, Saratov: Izdatelstvo saratovskogo universiteta.
Palterovich, D. and Moskvin, S. (1982) 'Sredstva maloi mekhanisatsii dlya selskogo khozyaistva', *Voprosy Ekonomiki*, no. 8.
Pauw, John W. de (1969) 'The private sector in Soviet agriculture', *Slavic Review* 28, no. 1.
Peredovaya (1982) 'Peredovaya', *Kommunist*, no. 11.
Pospielovsky, Dimitry (1970) 'The "Link System" in Soviet agriculture', *Soviet Studies* 21, no. 4.
Pravovoi (1984) *Pravovoi rezhim zemel v SSSR*, Moscow: Nauka.
Primernoe (1986) 'Primernoe polozhenie o podsobnom selskom khozyaistve, yavlyayushchemsya strukturnym podrazdeleniem predpriyatiya, organizatsii, uchrezhdeniya', *Ekonomika selskogo khozyaistva*, no. 5.
Primernyi (1970) *Primernyi ustav kolkhoza*, Moscow: Kolos.
Prodovolstvennaya (1982) *Prodovolstvennaya programma SSSR na period do 1990 goda i mery po ee realizatsii. Materialy maiskogo plenuma TsK KPSS*, Moscow: Politizdat.

Raig, Ivar Kh. (1984) 'Razvitie lichnogo podsobnogo khozyaistva v sovetskom derevne', *Istoriya SSSR*, no. 5.
Raig, Ivar Kh. (1986) 'Chto mozhet individualnoe khozyaistvo?', *Sotsiologicheskie Issledovania*, no. 1.
Rapoport, Anatol (1982) 'Prisoner's dilemma - recollections and observations', in Barry, Brian & Hardin, Russell (eds), *Rational Man and Irrational Society*, Beverly Hills, London and New Delhi: Sage.
Robinson, Geroid T. (1961) *Rural Russia Under the Old Regime*, London and New York: Longman.
Rutkevich, Mikhail N. (1985) 'O razvitii sovetskogo obshchestva k besklassovoi strukture', *Kommunist*, no. 18.
Rutkevich, Mikhail N. and Filippov, F. R. (1970) *Sotsialnye peremeshcheniya*, Moscow.
Schinke, Eberhard (1972) 'Soviet agricultural statistics', in Vladimir Treml and John Hardt (eds) *Soviet Economic Statistics*, Durham, NC: Duke University Press.
Selskokhozyaistvennaya (1972) *Selskokhozyaistvennaya entsiklopediya*, vol. III, Moscow: Sovetskaya entsiklopediya.
Selskokhozyaistvennaya (1973) *Selskokhozyaistvennaya entsiklopediya*, vol. IV, Moscow: Sovetskaya entsiklopediya.
Sergeev, Sergei S. (1956) *Voprosy ekonomiko-statistichesko analiza kolkhoznogo proizvodstva*, Moscow: Selkhozgiz.
Severin, Barbara S. (1979) 'USSR: the All-Union and Moscow collective farm market price indexes', *ACES Bulletin* 21, no. 1.
Severin, Barbara S. (1987) 'Solving the Soviet livestock feed dilemma: key to meeting Food Program targets', in US Congress, Joint Economic Committee, *Gorbachev's Economic Plans*, vol. II, Washington, DC: Government Printing Office.
Shenfield, Stephen (1984) 'How reliable are published Soviet data on the *kolkhoz* markets?', *CREES Discussion Papers*, General Series, G1, November.
Shmelev, Gelii I. (1964) *Raspredelenie i ispolzovanie truda v kolkhozakh*, Moscow: Mysl.
Shmelev, Gelii I. (1981) 'Obshchestvennoe proizvodstvo i lichnoe podsobnoe khozyaistvo', *Voprosy Ekonomiki* 53, no. 5.
Shmelev, Gelii I. (1983) *Lichnoe podsobnoe khozyaistvo: vozmozhnosti i perspektivy*, Moscow: Politicheskaya Literatura.
Shmelev, Gelii I. (1984) *Podsobnye khozyaistva predpriyatii i naseleniya*, Moscow: Znanie.
Shmelev, Gelii I. (1985) 'Lichnoe podsobnoe khozyaistvo', *Novoe v zhizni, nauke i tekhnike, Seriya Ekonomika*, no. 3.
Shmelev, Gelii I. (1987) *Semya beret podryad*, Moscow: Agropromizdat.
Shmelev, Nikolai (1987) 'Avansy i dolgi', *Novyi Mir*, no. 6.
Shmelev, Nikolai (1988) 'Novye trevogi', *Novyi Mir*, no. 4.
Sidorenko, V. I. (1985) 'Vazhnyi istochnik popolneniya prodovolstvennogo fonda', *Planirovanie i uchet v selskokhozyaistvennykh predpriyatiyakh*, no. 3.
Simis, Konstantin (1982) *USSR: The Corrupt Society; The Secret World of*

Soviet Capitalism, New York: Simon and Schuster.
Sobranie (1987) *Sobranie postanovlenii pravitelstva SSSR*, no. 10.
SSSR (1987) *SSSR v tsifrakh v 1986 g.*, Moscow: Finansy i statistika.
Stalin, Joseph V. (1952) *Ekonomicheskie problemy sotsializma v SSSR*, Moscow: Gospolitizdat.
Suslov, I. (1982) 'Kolkhozy v sisteme narodnogo khozyaistva', *Voprosy Ekonomiki*, no. 12.
Tenson, Andreas (1983) 'Factory farms or farm factories?', *Radio Liberty Research*, RL 71/83, 8 February.
Tenson, Andreas (1987) 'Easing of restrictions on private plots', *Radio Liberty Research*, RL 369/87, 18 September.
Thorner, David, Kerblay, Basile and Smith, R. E. F. (eds) (1966) *A. V. Chayanov on the Theory of Peasant Economy*, Homewood, IL: Irwin.
Tikhonov, I. (1982) 'Edinyi nardnokhozyaistvennyi kompleks mnogonatsionalnogo sovetskogo gosudarstva', *Kommunist*, no. 11.
Timofeev, Lev (1985) *Soviet Peasants (or: The Peasant's Art of Starving)*, New York: Telos.
Tipovoi (1981) 'Tipovoi ustav sadovodcheskogo tovarishchestva rabochikh i sluzhashchikh', in A.F. Kalinkin (ed.) *Lichnoe podsobnoe khozaistvo, kollektivnoe sadovodstvo i ogorodnichestvo*, Moscow: Kolos.
Treml, Vladimir D. (1985) 'Purchases of food from private sources in Soviet urban areas', *Berkeley-Duke Occasional Papers on the Second Economy in the USSR*, no. 3, September.
US Congress (1982) Joint Economic Committee, *Soviet Economy in the 1980s: Problems and Prospects*, vol. II, Washington, DC: Government Printing Office.
US Congress (1987) Joint Economic Committee, *Gorbachev's Economic Plans*, vol. II, Washington, DC: Government Printing Office.
USDA (1986) US Department of Agriculture, *USSR: Situation and Outlook Report*, Washington, DC: Government Printing Office.
Ushakov, V. M. (1981) 'Finansirovanie i kreditovanie grazhdan na razvitie lichnykh podsobnykh khozyaistv i zhilizhnoe stroitelstve na sele', in A.F. Kalinkin (ed.) *Lichnoe podsobnoe khozyaistvo, kollektivnoe sadovodstvo i ogorodnichestvo*, Moscow: Kolos.
Vasilev, Ivan (1982) 'Vozvrashchenie k zemle', *Nash sovremennik*, no. 6.
Volin, Lazar (1970) *A Century of Russian Agriculture*, Cambridge, MA: Harvard University Press.
Voronin, V. (1980) 'Lichnye podsobnye khozyaistva i torgovlya', *Voprosy Ekonomiki*, no. 6.
Wädekin, Karl-Eugen (1967) *Privatproduzenten in der sowjetischen Landwirtschaft*, Köln: Verlag Wissenschaft und Politik.
Wädekin, Karl-Eugen (1973) *The Private Sector in Soviet Agriculture*, Berkeley, CA: University of California Press.
Wädekin, Karl-Eugen (1983) 'The impact of official policy on the number of livestock in the Soviet private farming sector', *Radio Liberty Research*, RL 224/83, 9 June.
Wädekin, Karl-Eugen (1985) 'Private gardeners in the USSR', *Radio Liberty Research*, RL 174/85, 30 May.

Wädekin, Karl-Eugen (1985) 'The private agricultural sector in the 1980s', *Radio Liberty Research*, RL 251/85, 2 August.
Wädekin, Karl-Eugen (1985) 'The role of the *kolkhoz* market: a quantitative assessment', *Radio Liberty Research*, RL 155/85, 13 May.
Wädekin, Karl-Eugen (1986) 'Private Leistungen für den Lebensmittelsmarkt der UdSSR', *Osteuropa* 36, no. 1.
Wädekin, Karl-Eugen (1987) 'Agriculture', in M. McCauley (ed.) *The Soviet Union under Gorbachev*, New York: St Martin's.
Wädekin, Karl-Eugen (1987) 'Commentary', in US Congress, Joint Economic Committee, *Gorbachev's Economic Plans*, vol. II, Washington, DC: Government Printing Office.
Wädekin, Karl-Eugen (1988) 'The new *kolkhoz* statute: a codification of restructuring on the farm', *Radio Liberty Research*, RL 36/88, 28 January.
Wells, Herbert G. (1920) *Russia in the Shadows*, London: Hodder and Stoughton.
White, Stephen (1986) 'Economic performance and communist legitimacy', *World Politics* 38, no. 3.
Williamson, Oliver (1976) 'Some uses of the Exit–Voice approach – discussion', *American Economic Review* 66, no. 2.
Yanov, Alexander (1984) *The Drama of the Soviet 1960s. A Lost Reform*, Berkeley, CA: Institute of International Studies.
Zalygin, Sergei (1987) 'Povorot', *Novyi Mir*, no. 1.

Index of names

Subject Index

Note: All references are to Soviet private agriculture, except where otherwise specified.

For Product Safety Concerns and Information please contact our EU
representative GPSR@taylorandfrancis.com Taylor & Francis Verlag GmbH,
Kaufingerstraße 24, 80331 München, Germany

Printed and bound by CPI Group (UK) Ltd, Croydon, CR0 4YY
01/05/2025
01858464-0001